Biotechnology

T0179309

Biotechnology

Recent Trends and Emerging Dimensions

Edited by
Atul Bhargava
Shilpi Srivastava

CRC Press
Taylor & Francis Group
Boca Raton London New York

CRC Press is an imprint of the
Taylor & Francis Group, an **informa** business

CRC Press
Taylor & Francis Group
6000 Broken Sound Parkway NW, Suite 300
Boca Raton, FL 33487-2742

First issued in paperback 2020

© 2018 by Taylor & Francis Group, LLC
CRC Press is an imprint of Taylor & Francis Group, an Informa business

No claim to original U.S. Government works

ISBN 13: 978-0-367-57263-1 (pbk)
ISBN 13: 978-1-138-56108-3 (hbk)

Library of Congress Cataloging-in-Publication Data

Names: Bhargava, Atul, 1975- editor. | Srivastava, Shilpi, editor.
Title: Biotechnology : recent trends and emerging dimensions / editors: Atul Bhargava, Shilpi Srivastava.
Other titles: Biotechnology (Bhargava)
Description: Boca Raton : Taylor & Francis, 2018. | Includes bibliographical references and index.
Identifiers: LCCN 2017034225| ISBN 9781138561083 (hardback : alk. paper) | ISBN 9780203711033 (ebook)
Subjects: | MESH: Biomedical Technology | Biotechnology
Classification: LCC R855.3 | NLM W 82 | DDC 610.28--dc23
LC record available at https://lccn.loc.gov/2017034225

Visit the Taylor & Francis Web site at
http://www.taylorandfrancis.com

and the CRC Press Web site at
http://www.crcpress.com

Contents

Preface

Knowledge is twofold and consists not only in an affirmation of what is true, but in the negation of what is false.

The above-mentioned quote by the famous writer Charles Caleb Colton aptly describes the power of knowledge. It is in this pursuit of knowledge that this work was conceptualized.

Biotechnology has always been at the core of our heart. Ever since we started our academic journey, it was our inherent desire to introduce our students to the secrets of science and the power of knowledge; to explore the unknown and delve into the darkness in pursuit of light, that is, the light of knowledge. In this book, we have attempted to explore some of the emerging areas of biotechnology, which are bound to change the future of the human society in the decades to come.

Biotechnology has been an integral component of mankind. Biotechnology is defined as any technological application that uses biological systems, living organisms, or derivatives thereof, to make or modify products or processes for specific use. This field has advanced rapidly in the recent decades with enormous information available to us every passing day. It is one of the fastest emerging sectors in the world economy having a marked impact on almost all domains of human welfare, and the biotechnology market is expected to reach US $604.40 billion by the year 2020. The biotechnology realm encompassing the areas of microbiology, biochemistry, genetics, molecular biology, tissue engineering, immunology, agriculture, environment, and cell and tissue culture physiology, and, in some instances, is also dependent on knowledge and methods from several sciences outside the biological sphere such as chemical engineering, bioprocess engineering, information technology, and biorobotics. As biotechnology has the potential to affect human lives, both directly and indirectly, new developments in this field are bound to have a deep impact on human development. Looking at this multidisciplinary field and the tremendous scope of biotechnology, the time is ripe enough to bring out a book, which can guide the readers to these technologies in a simple way. It is in this respect that the present book has been conceptualized so that the reader may get a detailed view of the recent advances in this rapidly emerging field. This proposed book explains these recent developments in biotechnology along with their applications to the readers so that even a layman can have an idea of how the upcoming subfields of biotechnology can add to human welfare.

This proposed book describes the recent developments in the field of biotechnology in an easy-to-read, succinct format, and guides the readers to the latest updates in this ever-expanding field. This book addresses to what is new and significant. Spread over 10 chapters, this book gives an overview of the latest research and developments with an in-depth coverage of recent developments in the field. The topics are presented in a comprehensible, coherent, and a methodical way and care has been taken to keep the level of discussion simple and free from any technical jargon. The chapters are comprehensible, straightforward, with illustrations to make the topics simple and provide additional help and clarity. The authors have tried to present complex and updated information in a manner that familiarizes the reader with the important concepts and tools of recent biotechnological researches. Apart from core biotechnology students, this book would be useful for readers of diverse disciplines such as pharmaceutical sciences, medical sciences, nutritional science, and general biology.

We hope that this book will help emerging biotechnologists to understand the various facets of this fascinating science, which is bound to rule the world in the twenty-first century.

Atul Bhargava

Shilpi Srivastava

Acknowledgments

We would say that the students have been our greatest inspiration and guiding force in conceptualizing this work. It is the students, not just at Amity, but also around the world for whom this text has primarily been written.

We would also thank all our contributing authors who have worked tirelessly to produce the chapters of high excellence.

We are deeply indebted to our parent organization Amity University Uttar Pradesh, Noida, India and Dr. Aseem Chauhan, Chairman, Amity University Uttar Pradesh, Lucknow Campus, India, for providing us a suitable forum to fulfill our ambitions. We also thank Dr. J.K. Srivastava, head, Amity Institute of Biotechnology, Lucknow, India, for his constant support and encouragement.

It is a great opportunity for us to express our deep sense of gratitude and heartiest veneration to Dr. Deepak Ohri, deputy dean Research at Amity University Uttar Pradesh, Lucknow Campus, India, for his inspiring guidance and constant encouragement in completing this endeavor.

We thank our colleagues and friends at the Amity University: Dr. Prachi Srivastava, Dr. Rachna Chaturvedi, Dr. Sonia Chadha, Dr. Priti Mathur, Dr. Jyoti Prakash, and Ms. Garima Awasthi for their support and psychostimulant company.

The acknowledgment would be incomplete without thanks for the immense cooperation of Dr. Anand Vardhan Srivastava for his wholehearted support and for always being with us in the hour of need.

Special thanks and appreciation are due to Dr. Vijai Kumar Gupta, Tallinn University of Technology, Tallinn, Estonia and Dr. Nitin Kumar Sharma for their wholehearted help, advice, and encouragement in our present endeavor.

On a very personal note, we are grateful to our brother Mr. Akhilesh Bhargava and niece Ms. Anushka Sharma for their patience and perseverance, as well as their constant support during the writing of this book. They wholeheartedly supported in our overburdened schedule and were with us during the thick and thin.

We express our appreciation for the editorial guidance of Dr. Renu Upadhyay (Commissioning Editor, CRC Press) and Ms. Shikha Garg (Editorial Assistant, CRC Press) whose efforts have helped us in shaping the text to bring out the best to the readers.

A final word of gratitude to all those who did not find mention above but have contributed somehow in the publication of this book.

Atul Bhargava

Shilpi Srivastava

Contributors

Vineet Awasthi
Amity Institute of Biotechnology
Amity University
Lucknow, India

Atul Bhargava
Amity Institute of Biotechnology
Amity University
Lucknow, India

Meenakshi Bhargava
Department of Chemistry
Central University of Allahabad
Allahabad, India

Francisco Fuentes
Facultad de Agronomía e Ingeniería
    Forestal Pontificia
Universidad Católica de Chile
Santiago, Chile

Gurjeet Kaur
Amity Institute of Biotechnology
Amity University
Lucknow, India

Arushi Misra
Department of Biotechnology
Kumaun University
Bhimtal, India

Ajay Kumar Singh
Center for Biological Sciences
    (Bioinformatics)
School of Earth, Biological and
    Environmental Science
Central University of South Bihar
Patna, India

Vandana Singh
Department of Chemistry
Central University of Allahabad
Allahabad, India

Shilpi Srivastava
Amity Institute of Biotechnology
Amity University
Lucknow, India

1 Biosorption

Atul Bhargava

CONTENTS

1.1 INTRODUCTION

The rapid industrialization, urbanization, and modern agricultural practices have adversely affected the ecosystem and generated large quantities of aqueous effluents, many of which contain high levels of toxic pollutants. Human activities and establishment of various industries such as tanneries, textile plants, chemical works, electrolysis, electro-osmosis, metal fabrication shops, paper mills, waste disposal sites, and intensive agriculture have contributed a lot in polluting the ecosystems. The contamination of the environment with toxicants has become a worldwide problem that affects crop yields, soil biomass, and fertility and leads to bioaccumulation of these toxicants in the food chain. Aquatic bodies are increasingly being overwhelmed by aqueous effluents containing large amounts of toxic pollutants such as heavy metals and dyes. The dyes and color pigments also contain heavy metals such as chromium, copper, nickel, mercury, and cobalt. The toxic substances have a tendency to accumulate in the food chain and induce toxicity-related problems, not only for the environment but also for human beings. Increasing environmental pollution from industrial wastewater is of major concern, particularly in the developing countries. Several countries, especially the developed ones, have regulated the emission of toxic substances, but in the third world countries, intense industrial development and population increase, coupled with lack of pollution control measures, have caused an enormous increase in heavy metal contamination of agricultural soils and water bodies. The organic and inorganic pollutants exert a detrimental effect on the already-fragile ecosystems and on human health. Therefore, a number of physical, chemical, and biological methods are resorted to prevent or limit industrial discharges, which incur considerable expenditure.

1

The remarkable features of living organisms in the transformation and detoxification of different pollutants have been well documented and form an important component of environmental biotechnology and microbiology. The use of biosorbents for the removal and recovery of toxic pollutants from aquatic ecosystems is one of the most recent developments in environmental biotechnology.

1.2 WHAT IS BIOSORPTION?

Adsorption is defined as the physical adherence or bonding of ions and molecules onto the surface of the solid material. Biosorption is a subcategory of adsorption, where the sorbent is a biological matrix. Thus, biosorption can be defined as the removal of metal or metalloid species, compounds, and particulates from solution by biological material. Bohumil Volesky, a professor at McGill University, has defined biosorption as the property of certain biomolecules (or types of biomass) to bind and concentrate selected ions or other molecules from aqueous solutions. The process makes use of either living or inexpensive dead biomass to sequester toxic substances and is of immense use in the removal of these contaminants from industrial effluents. Biosorption involves reversible binding of ions from solutions onto the functional groups present biomass surface. The process is quite rapid and is independent of cellular metabolism. Biosorption capacity of a substance often varies with test conditions such as initial concentration of the contaminant, solution pH, biomass dosage, contact time, and the processing method. Biosorption can be performed in a wide range of pH (3–9) and temperature values (4°C–90°C). pH is one of the key determinants that influences solution chemistry of metal ions, hydrolysis, dissociation of sites, complexation by organic and/or inorganic ligands, precipitation, redox reactions, speciation, and the biosorption affinity of metal ions. Since the optimum particle size of the biosorbent is quite small (1–2 mm), the equilibrium state of both adsorption and desorption is achieved rapidly. Table 1.1 depicts the

TABLE 1.1
A Comparison of Biosorption and Bioaccumulation

Feature	Biosorption	Bioaccumulation
Property	Passive process	Active process
Location	Extracellular	Intra- and intercellular
Temperature	Temperature-independent	Temperature-dependent
Rate of uptake	Fast	Slow
Cost	Low	High
Sensitivity	Nutrient-independent	Nutrient-dependent
Selectivity	Less	More
Regeneration and use	High possibility	Less possibility
Commercial applicability	More	Less

major differences between biosorption and bioaccumulation, both of which are concerned with the removal of contaminants from aquatic systems.

1.3 BRIEF HISTORY

In 1902, it was Hecker who first reported a quantitative study on the copper uptake by fungal spores of *Tilletia tritici* (the causal agent of common bunt of wheat) and *Ustilago crameri*. In 1922, F. Pichler and A. Wobler reported the uptake of silver (Ag), mercury (Hg), copper (Cu), and cerium (Ce) by corn smut. Efficient removal of radioactive metals such as plutonium-239 from contaminated domestic sewage by activated sludge was reported in 1949 by Ruchloft. The use of biosorption using mosses for monitoring the presence of trace elements in the environment was first reported in 1971 by Goodman and Roberts. Neufeld and Hermann (1975) studied the kinetics of biosorption by activated sludge and found initial rapid uptake of heavy metals such as cadmium (Cd), mercury (Hg), and zinc (Zn), followed by a slow uptake over the next few hours. Friedman and Dugan (1968) used a pure culture of *Zoogloea*, a gram-negative, aerobic, rod-shaped bacteria for assessing the accumulation of metallic ions. Sakaguchi et al. (1978) and Nakajima et al. (1982) reported that the ability of microorganisms to accumulate uranium ions was in the order of actinomycetes > bacteria > yeast > fungi. In 1982, B. Volesky and M. Tsezos were awarded the first patent on the use of biosorption technology for removing uranium and thorium ions from the aqueous suspension or solution. Uranium and thorium cations were removed from the aqueous suspension or solution by treatment of the aqueous material with the biomass derived from fermentation of the fungal genus *Rhizopus*. They proposed that the process could be utilized to treat aqueous tailings from uranium ore extraction processes for reducing the radioactive content of the tailings before disposal. The competition between metal ions for the binding sites of anaerobic sludge was evaluated by Gould and Genetelli (1984), who reported a binding affinity order of Cu > Cd > Zn > Ni.

1.4 BIOSORBENTS

The initial step in biosorption was the selection of the most suitable types of biomass from a large reservoir of readily available and inexpensive biomaterials. Availability and cost are the main criteria taken into consideration when selecting biomass for large-scale industrial use. A large pool of biomaterials available in nature can be employed as biosorbents for the removal of the desired pollutant. Biosorbents are prepared from the naturally abundant and/or waste biomass of living forms such as algae, moss, fungi, and even bacteria. Bacteria are known to make excellent biosorbents because of their high surface-to-volume ratios and significant amount of potentially active chemosorption sites such as teichoic acid as component in their cell walls. However, bacteria and fungi need to be cultivated, which increases the cost, and the cultivation practices increase the uncertainty in maintaining a continuous supply of biomass for the process. Seaweeds, owing to their rigid macroscopic structures, offer a convenient basis for the production of biosorbent particles suitable for sorption applications. Seaweeds are abundant, show rapid growth, and at places threaten the tourist industry by spoiling pristine environments and giving rise to fouling beaches. However, seaweeds are economically very important in being a major source for the production of industrial products such as agar, alginate, and carrageenan. Therefore, utmost care should be taken while selecting seaweeds for the biosorption process.

Apart from the natural biosorbents, biomaterials such as plant barks, leaves, rice husk, black gram husk, potato peels, coconut shell, egg shells, seed shells, sawdust, citrus peels, sugarcane bagasse, peat moss, and fly ash have also generated much interest recently. Figure 1.1 provides details of the various kinds of biosorbents used in the biosorption process. It has been demonstrated that biosorptive processes using nonliving biomass are in fact more feasible than the bioaccumulative processes that use living organisms, since the latter require a constant nutrient supply and complicated bioreactor systems. The dead biomass exhibit specific advantages in comparison with the use of living organisms, which are enumerated as follows:

1. The dead biomass can be stored with ease for long durations.
2. Dead cells are not subject to the limitations of metal toxicity.
3. There is no requirement of nutrient supply.
4. Metal ion-loaded biosorbents display easy desorption and reuse.

However, the use of dead biomass has limitations such as small particle size, low mechanical strength, cumbersome process for biomass separation from the reaction system, and mass loss after regeneration, which limits its use in batch and continuous systems.

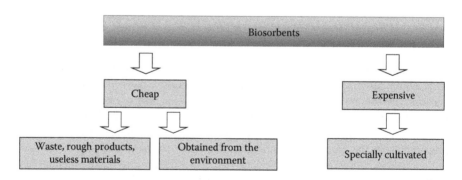

FIGURE 1.1 Economic considerations for different kinds of biosorbents used in the biosorption process. (Adapted from Michalak, I. et al., *Appl. Biochem. Biotechnol.*, 170, 1389–1416, 2013.)

1.5 MECHANISM OF BIOSORPTION

The mechanism of biosorption is complex and can be broadly divided into the following two types:

1. Physical sorption: This is caused by van der Walls forces or Coulomb forces. Since biosorbents have large values of the Langmuir and the Brunauer–Emmett–Teller (BET) surface area, the contaminant can accumulate on its surface.
2. Chemical sorption: It is caused by complexation, ion exchange, coordination, or oxidation–reduction.

However, researchers have also divided biosorption based on other criteria. Another way of differentiating is based on the dependence on the cell's metabolism, whereby biosorption can be divided into the two types:

1. Metabolism-dependent
2. Nonmetabolism-dependent

Ahalya et al. (2003) has divided biosorption into the following three types, based on the location where the contaminant removed from solution is found:

1. Extracellular accumulation/precipitation
2. Cell surface sorption/precipitation
3. Intracellular accumulation

Some of the above-mentioned mechanisms may occur simultaneously due to intricate nature of the biosorbents.

1.6 FACTORS AFFECTING BIOSORPTION

The biosorption process is complex and is influenced by a number of factors. These are as follows:

1. pH is recognized as the most important parameter in the biosorptive process. pH affects the activity of the functional groups in the biomass, the solution chemistry of metals, and the competition of metallic ions.
2. Biomass concentration in solution is also known to influence the specific uptake. There is an increase in the specific uptake for lower values of biomass concentrations. The initial concept that an increase in biomass concentration interferes with the binding sites was later modified by attributing the responsibility of the decrease in specific uptake of the metal to the reduction in metal concentration in the solution. Therefore, biomass concentration should be considered while using microbial biomass as biosorbent.

3. Biosorption is also influenced by the presence of other metal ions apart from the target ions present in the solution. For example, the presence of zinc and iron influences uranium uptake by *Rhizopus arrhizus*. Likewise, cobalt uptake by different microbes seems to be inhibited by the presence of lead, mercury, uranium, and copper.

1.7 APPLICATIONS

Biosorption has been utilized to remove a wide range of toxic substances from the aqueous medium. These substances include synthetic dyes, heavy metals, fluoride, phthalates, and pharmaceuticals. However, removal of dyes (Table 1.2), heavy metal ions, metalloids, actinides, lanthanides, and radioisotopes has been at the forefront of biosorption research (Tables 1.3 through 1.6).

An example of the use of biosorbents for heavy metal remediation is provided later. The biomass is first inactivated and then pretreated by washing with acids and/or bases before final drying and granulation. Although cutting and/or grinding of the dry biomass usually yields stable biosorbent particles, some types have to be immobilized in a synthetic polymer matrix and/or grafted on an inorganic support material to yield particles with the requisite mechanical properties. The biosorbent particles are then packed in sorption columns, which operate on cycles comprising various steps such as loading, regeneration, and rinsing. These columns have been most effectively utilized for continuous removal of contaminants. The operation

TABLE 1.2
Various Life Forms as Biosorbents for Remediation of Dyes

Scientific Name	Dye	Reference
Fomes fomentarius, Phellinus igniarius	Methylene blue and Rhodamine B	Maurya et al. (2006)
Rhizopus arrhizus	Methylene blue	Aksu et al. (2010)
Aspergillus flavus	Remazol black B	Ranjusha et al. (2010)
Pseudomonas putida	Direct red	Deepaa et al. (2013)
Azolla filiculoides	Redwine dye	Rani et al. (2013)
Salvinia minima	Methylene blue	Sánchez-Galván and Ramírez-Núñez (2014)
Sphagnum magellanicum	Basic blue 3 and Basic orange	Contreras et al. (2007)
Stoechospermum marginatum	Acid blue 25, Acid orange 7 and Acid black 1	Daneshvar et al. (2012)
Posidonia oceanica	Methylene blue	Ncibi et al. (2007)
Cyperus rotundus	Crystal violet	Suyamboo and Srikrishnaperumal (2014)
Paulownia tomentosa	Acid orange 52	Deniz and Saygideger (2010)
Corynebacterium glutamicum	Reactive black 5	Vijayaraghavan and Yun (2007)
Capsicum annuum	Reactive blue 221	Gürel (2017)

TABLE 1.3
Bacteria as Biosorbents for Remediation of Metals

Bacteria	Metal	Reference
Aeromonas caviae	Cadmium, chromium	Loukidou et al. (2004)
Aphanothece halophytica	Zinc	Incharoensakdi and Kitjaharn (2002)
Arthrobacter nicotianae	Uranium	Nakajima and Tsuruta (2004)
Bacillus firmus	Copper, lead	Salehizadeh and Shojaosadati (2003)
Bacillus licheniformis	Chromium	Zhou et al. (2007)
Bacillus subtilis	Copper	Nakajima et al. (2001)
Bacillus thuringiensis	Nickel	Ozturk (2007)
Corynebacterium glutamicum	Lead	Choi and Yun (2004)
Desulfovibrio desulfuricans	Palladium, platinum	de Vargas et al. (2004)
Enterobacter sp.	Lead, cadmium	Lu et al. (2006)
Micrococcus luteus	Copper	Nakajima et al. (2001)
	Uranium	Nakajima and Tsuruta (2004)
Nocardia erythropolis	Uranium	Nakajima and Tsuruta (2004)
Ochrobactrum anthropi	Cadmium	Ozdemir et al. (2003)
Pseudomonas aeruginosa	Cadmium	Chang et al. (1997)
Pseudomonas putida	Lead	Uslu and Tanyol (2006)
Streptomyces rimosus	Zinc	Mameri et al. (1999)
	Lead	Selatnia et al. (2004)
Sphaerotilus natans	Copper	Beolchini et al. (2006)
Sphingomonas paucimobilis	Cadmium	Tangaromsuk et al. (2002)
Thiobacillus ferrooxidans	Zinc, copper	Liu et al. (2004)
Zoogloea ramigera	Chromium	Nourbakhsh et al. (1994)
	Uranium	Nakajima and Tsuruta (2004)

TABLE 1.4
Fungi as Biosorbents for Remediation of Metals

Fungi	Metal	Reference
Pseudomonas aeruginosa	Uranium	Strandberg et al. (1981)
Saccharomyces cerevisiae	Cadmium	Volesky et al. (1993)
Aspergillus niger	Cadmium	Barros Júnior et al. (2003)
Aspergillus niger	Lead	Eram et al. (2015)
Aspergillus flavus	Copper	Eram et al. (2015)
Aspergillus oryzae	Chromium	Nasseri et al. (2002)
Mucor rouxii	Cadmium, nickel, lead	Yan and Viraraghavan (2008)
Talaromyces helices	Copper	Romero et al. (2006)
Penicillium purpurogenum	Chromium	Say et al. (2004)
Agaricus macrosporus	Chromium, lead, zinc	Melgar et al. (2007)
Penicillium simplicissimum	Lead, copper	Xiao-Ming et al. (2008)
Phanerochaete chrysosporium	Cadmium, lead	Day et al. (2001)
Rhizopus arrhizus	Cadmium, zinc	Kuyucak and Volesky (1988)

TABLE 1.5
Biosorption of Heavy Metals by Lower Plant Forms

Plant Forms	Biological Name	Heavy Metal	Reference
Algae	*Chaetomorpha linum*	Copper, zinc	Ajjabi and Chouba (2009)
	Caulerpa lentillifera	Cadmium, lead	Pavasant et al. (2006)
	Cladophora fascicularis	Lead	Liping et al. (2007)
	Rhodotorula glutinis	Lead	Cho and Kim (2003)
	Ceramium virgatum	Cadmium	Ahmet and Mustafa (2008)
	Fucus vesiculosus	Cadmium	Brinza et al. (2009)
	Spirulina sp.	Chromium	Chojnacka et al. (2005)
	Ulva reticulata	Nickel	Vijayaraghavan et al. (2005)
	Ulva lactuca	Cadmium	Asnaoui et al. (2015)
	Galdieria sulphuraria	Palladium, gold	Ju et al. (2016)
	Durvillaea antarctica	Cadmium	Gutiérrez et al. (2015)
	Chlorella minutissima	Cadmium, copper	Yang et al. (2014)
	Scenedesmus spinosus	Strontium,	Liu et al. (2014)
	Laminaria japonica	Uranium	Lee et al. (2014)
Bryophytes	*Racomitrium lanuginosum*	Palladium	Sari et al. (2009)
	Drepanocladus revolvens	Mercury	Sari and Tuzen (2009)
	Fontinalis antipyretica	Lead	Martins and Boaventura (2011)
	Fontinalis antipyretica	Zinc	Martins and Boaventura (2002)
	Sphagnum sp.	Nickel	Ho et al. (1995)
Pteridophytes	*Acrostichum aureum*	Zinc, lead	Lobo Soniya and Gulimane (2015)
	Salvinia sp.	Nickel, cobalt	Dhir et al. (2010)
	Athyrium filix-femina	Zinc, copper	Asiagwu et al. (2012)
	Azolla pinnata	Lead, copper	El-All et al. (2011)
	Azolla filiculoides	Gold	Umali et al. (2006)
		Strontium	Mashkani and Ghazvini (2009)

TABLE 1.6
Angiosperms as Biosorbents for Remediation of Heavy Metals

Angiosperms	Heavy Metal	Reference
Aegle marmelos	Chromium	Anandkumar and Mandal (2009)
	Lead	Chakravarty et al. (2010)
Avena monida	Chromium	Gardea-Torresdey et al. (2000)
Azadirachta indica	Zinc	King et al. (2007)
Borassus aethiopum	Chromium	Elangovan et al. (2008)
Callitriche cophocarpa	Chromium	Augustynowicz et al. (2010)
Cassia fistula	Nickel	Hanif et al. (2007)
Ceratophyllum demersum	Nickel	Chorom et al. (2012)
Cicer arietinum	Cadmium	Saeed and Iqbal (2003)
Citrus reticulata	Nickel	Ajmal et al. (2000)
Eichhornia crassipes	Chromium, zinc	Mishra and Tripathi (2009)
Hevea brasiliensis	Chromium	Karthikeyan et al. (2005)
Hydrilla verticillata	Cadmium	Bunluesin et al. (2007)
Hypnea valentiae	Cadmium	Aravindhan et al. (2009)
Lathyrus sativus	Cadmium	Panda et al. (2006)
Madhuca longifolia	Lead	Rehman et al. (2013)
Manihot esculenta	Cadmium, copper, zinc	Horsfall et al. (2006)
Ocimum americanum	Chromium	Lakshmanraj et al. (2009)
Parthenium hysterophorus	Cadmium	Ajmal et al. (2006)
Platanus orientalis	Cadmium	Mahvi et al. (2007)
Solanum nigrum	Cadmium	Luo et al. (2011)
Syzygium jambolanum	Mercury, chromium	Muthukumaran and Sophie Beulah (2010)
Tectona grandis	Copper	King et al. (2006)
Zea mays	Lead, chromium	Garcia-Rosales et al. (2012)

is initiated by loading the sorbent material and passing the metal-bearing effluent through the packed bed. In this process, the contaminants such as heavy metals are taken up from the liquid by the biosorbent. After exhaustion of the metal sorption capacity of the biosorbent, the column is taken out of the unit, and its bed is regenerated with solutions of acids and/or hydroxides. The rinsing and/or backwashing of the bed with water to remove the remains of the regenerants and suspended solids captured in the column mark the culmination of the cycle. Two columns are simultaneously employed to make the biosorption process truly continuous, wherein one is being regenerated and rinsed, and the other is being loaded with heavy metals.

1.8 ECONOMIC FEASIBILITY

In the United States and Canada, pilot installations and commercial units constructed and operated during the 1980s and 1990s have confirmed that biosorption can be effectively used for metal sequestering and the recovery processes, especially in the case of uranium. The *in situ* biological leaching of uranium at Denison Mines in Elliott Lake District of Canada has resulted in significant cost reduction as compared with conventional methods. With reference to metal sorption, biosorption is a process with some unique characteristics. Biosorption is considered an ideal method for the treatment of large volumes of low-concentration complex wastewaters, since the dissolved metals can be effectively sequestered from very dilute solutions with great efficiency. Biosorption, together with other techniques such as bioprecipitation and bioreduction, contributes as a parallel mechanism in cases of metabolically active microbial cells in biological reactors. Thus, biosorption should be regarded as a metal-immobilization technique in metal-bearing water treatment technology, which is based on the interactions of microbial cells with soluble metal species.

1.9 ADVANTAGES

Kratchovil and Volesky (1998) and Sud et al. (2008) have discussed numerous advantages of biosorption over conventional treatment methods. These include the following:

1. Low cost of biosorbents, since they are made from abundant or waste material.
2. High efficiency of contaminant removal at low concentration. In fact, metal concentration has been reduced to a very low level, such as drinking water standards in some cases.
3. No secondary problem by minimization of chemical and biological sludge as compared with other techniques such as precipitation.
4. Regeneration of biosorbent after removal of the contaminant.
5. In case of removal of metals, there is a possibility of recovery of metals after being sorbed from the solution. There is also a great advantage of selectivity for heavy metals over alkaline earth metals during the biosorption process.
6. More environment-friendly life cycle of the material.
7. Versatility and flexibility for a wide range of applications.
8. No nutrient requirement.

1.10 FUTURE PERSPECTIVES

Biosorption has proven to be a potential cost-effective technology for the treatment of high-volume low-concentration complex wastewaters contaminated with metal ions. Efforts are on to identify natural biosorbents and to assess their compatibility for industrial effluents. However, attempts to scale up the biosorption process or to commercialize it as a technology based on experiences from conventional sorption experiments in the laboratory have been limited. Despite a large number of publications on biosorption in the last few decades, the transfer of knowledge from laboratory to the industry has been a slow process. Most of this research on biosorption has been carried out using batch tank reactors or packed minicolumns in a laboratory setup, and very limited examples of industrial processes or products in this area are available. Table 1.7 provides the details of several commercial biosorbents

TABLE 1.7
Some Commercially Available Biosorbents

Biosorbent	Company	Organism as Biosorbent	Application
BIO–FIX™	U.S. Bureau of Mines (Golden, CO, U.S.)	*Spirulina, Lemna, Sphagnum*	Removal of heavy metal ions rom industrial wastewaters, acid mine drainage (AMD) waters and ground waters
AMT–BIOCLAIM™	Advanced Mineral Technologies, Inc. (AMT)	*Bacillus subtilis*	Removal of heavy metal ions and recovery of precious metals
AlgaSORB™	Bio-Recovery System, Inc. (Las Cruces, U.S.)	*Chlorella vulgaris*	Removal of heavy metal ions
BV–SORBEX™	BV SORBEX, Inc. (Montreal, Canada)	*Sphaerotilus natans, Ascophyllum nodosum, Halimeda opuntia, Palmyra pamata, Chondrus crispus, Chlorella vulgaris*	Removal of metal ions
MetaGeneR	–	–	Removal of heavy metal ions from wastewaters
Tsezos–Baird–Shemilt	–	*Rhizopus arrhizus*	Removal of metal ions
Bio-Beads™	RAHCO	Peat moss	Removal of heavy metal ions from wastewaters

that have been developed. Metal biosorption by synthetic or biosynthetic chemicals has also been investigated, wherein a mercury-binding synthetic biosorbent called Vitrokele™ 573 (comprising Hg covalently fixed to the surface of a suitable insoluble carrier) was prepared and used for removal of mercury. The column and batch tests demonstrated the effective removal of mercury and the reusage of the biosorbent over multiple cycles. In another case, an iron-binding biosorbent of synthetic nature was tested in a column containing radioactive iron, cobalt, cadmium, and sodium. The results showed good affinity for iron by the biosorbent, poor affinity for cobalt, and none for cadmium or sodium.

Biosorption is still in its developmental stages, and further improvement in both performance and costs can be expected. For removal of inorganics from industrial effluents, biosorption needs to effectively compete on both cost and performance bases with the existing methods, before industry accepts and implement it. The application of biosorption in case of metals exhibits some advantages such as feasibility at low metal concentration; effectiveness over a broad range of conditions that include a broad range of pH (3–9) and temperature (4°C–90°C); low capital investment and low operation cost; and conversion of pollutant metals to a metal product, thus eliminating the cost and liability to dispose of toxic sludge. Biosorption might be important where heavy metals need to be extracted from relatively dilute solutions such as laboratory waste effluent, in which metal ions are present in microgram to milligram levels. The immense potential of biosorption, coupled with its strong economic and technical advantages, opens up considerable market opportunities, but a lot of work needs to be done before the actual launching and commercialization of the biosorption technology.

BIBLIOGRAPHY

Abdolali A., Guo W.S., Ngo H.H., Chen S.S., Nguyen N.C. and Tung K.L. 2014. Typical lignocellulosic wastes and by-products for biosorption process in water and wastewater treatment: A critical review. *Bioresour. Technol.* 160: 57–66.

Ahalya N., Ramachandra T.V. and Kanamadi R.D. 2003. Biosorption of heavy metals. *Res. J. Chem. Environ.* 7: 71–79.

Ahmet S. and Mustafa T. 2008. Biosorption of cadmium (II) from aqueous solution by red algae (*Ceramium virgatum*): Equilibrium, kinetic and thermodynamic studies. *J. Hazard. Mater.* 157: 448–454.

Ajjabi L.C. and Chouba L. 2009. Biosorption of Cu^{2+} and Zn^{2+} from aqueous solutions by dried marine green macroalga *Chaetomorpha linum*. *J. Environ. Manag.* 90: 3485–3489.

Ajmal M., Rao R.A.K., Ahmad R. and Ahmad J. 2000. Adsorption studies on *Citrus reticulata* (fruit peel of orange): Removal and recovery of Ni (II) from electroplating wastewater. *J. Hazard. Mater.* 79: 117–131.

Ajmal M., Rao R.A.K., Ahmad R. and Khan M.A. 2006. Adsorption studies on *Parthenium hysterophrous* weed: Removal and recovery of Cd (II) from wastewater. *J. Hazard. Mater.* 135: 242–248.

Aksu Z., Ertuğrul S. and Dönmez G. 2010. Methylene blue biosorption by *Rhizopus arrhizus*: Effect of SDS (sodium dodecylsulfate) surfactant on biosorption properties. *Chem. Eng. J.* 158: 474–481.

Anandkumar J. and Mandal B. 2009. Removal of Cr (VI) from aqueous solution using Bael fruit (*Aegle marmelos correa*) shell as an adsorbent. *J. Hazard. Mater.* 168: 633–640.

Aravindhan R., Bhaswant M., Sreeram K.J., Raghava R.J. and Balachandran U.N. 2009. Biosorption of cadmium metal ion from simulated wastewaters using *Hypnea valentiae* biomass: A kinetic and thermodynamic study. *Biores. Technol.* 101: 1466–1470.

Arief V.O., Trilestari K., Sunarso J., Indraswati N. and Ismadji S. 2008. Recent progress on biosorption of heavy metals from liquids using low cost biosorbents: Characterization, biosorption parameters and mechanism studies. *Clean* 36: 937–962.

Asiagwu A.K., Owamah I.H. and Otutu J.O. 2012. Kinetic model for the sorption of Cu (II) and Zn (II) using lady fern (*Athyrium-Filix–Femina*) leaf waste biomass from aqueous solution. *Chem. Process. Eng. Res.* 3: 1–13.

Asnaoui H., Laaziri A. and Khalis M. 2015. Study of the kinetics and the adsorption isotherm of cadmium (II) from aqueous solution using green algae (*Ulva lactuca*) biomass. *Water Sci. Technol.* 72: 1505–1515.

Atkinson B.W., Bux F. and Kasan H.C. 1998. Considerations for application of biosorption technology to remediate metal-contaminated industrial effluents. *Water SA.* 24: 129–135.

Augustynowicz J., Grosicki M., Hanus-Fajerska E., Lekka M., Waloszek A. and Kołoczek H. 2010. Chromium (VI) bioremediation by aquatic macrophyte *Callitriche cophocarpa* Sendtn. *Chemosphere* 79: 1077–1083.

Barros Júnior L.M., Macedo G.R., Duarte M.M.L., Silva E.P. and Lobato A.K.C.L. 2003. Biosorption of cadmium using the fungus *Aspergillus niger*. *Braz. J. Chem. Eng.* 20: 229–239.

Beolchini F., Pagnanelli R., Toro L. and Veglio F. 2006. Ionic strength effect on copper biosorption by *Sphaerotilus natans*: Equilibrium study and dynamic modeling in membrane reactor. *Water Res.* 40: 144–152.

Bhargava A., Shukla S., Srivastava J., Singh N. and Ohri D. 2008. *Chenopodium*: A prospective plant for phytoextraction. *Acta Physiol. Plant.* 30: 111–120.

Brinza L., Nygard C.A., Dring M.J., Gavrilescu M. and Benning L.G. 2009. Cadmium tolerance and adsorption by the marine brown alga *Fucus vesiculosus* from the Irish Sea and the Bothnian Sea. *Biores. Technol.* 100: 1727–1733.

Bunluesin S., Kruatrachue M., Pokethitiyook P., Upatham S. and Lanza G.R. 2007. Batch and continuous packed column studies of cadmium biosorption by *Hydrilla verticillata* biomass. *J. Biosci. Bioeng.* 103: 509–513.

Chakravarty S., Mohanty A., Sudha T.N., Upadhyay A.K., Konar J., Sircar J.K., Madhukar A. and Gupta K.K. 2010. Removal of Pb (II) ions from aqueous solution by adsorption using bael leaves (*Aegle marmelos*). *J. Hazard. Mater.* 173: 502–509.

Chang J.S., Law R. and Chang C.C. 1997. Biosorption of lead, copper and cadmium by biomass of *Pseudomonas aeruginosa* PU21. *Water Res.* 31: 1651–1658.

Chiu Y., Asce M. and Zajic J.E. 1976. Biosorption isotherm for uranium recovery. *J. Environ. Eng. ASCE* 102: 1109–1111.

Cho D.H. and Kim E.Y. 2003. Characterization of Pb^{2+} biosorption from aqueous solution by *Rhodotorula glutinis*. *Bioproc. Biosystems Eng.* 25: 271–277.

Choi S.B. and Yun Y.S. 2004. Lead biosorption by waste biomass of *Corynebacterium glutamicum* generated from lysine fermentation process. *Biotechnol. Lett.* 26: 331–336.

Chojnacka K., Chojnacki A. and Górecka H. 2005. Biosorption of Cr^{3+}, Cd^{2+} and Cu^{2+} ions by blue-green algae *Spirulina* sp.: Kinetics, equilibrium and the mechanism of the process. *Chemosphere* 59: 75–84.

Chorom M., Parnian A. and Jaafarzadeh N. 2012. Nickel removal by the aquatic plant (*Ceratopyllum demersum* L.). *Intern. J. Environ. Sci. Dev.* 4: 1–4.

Chuah T.G., Jumasiah A., Azni I., Katayon S. and Choong S.Y.T. 2005. Rice husk as a potentially low-cost biosorbent for heavy metal and dye removal: An overview. *Desalination* 175: 305–316.

Contreras E., Martinez B., Sepulveda L. and Palma C. 2007. Kinetics of basic dye adsorption onto *Sphagnum magellanicum* Peat. *Adsorpt. Sci. Technol.* 25: 637–646.

Crini G. 2005. Recent developments in polysaccharide-based materials used as adsorbents in wastewater treatment. *Prog. Poly. Sci.* 30: 38–70.

Daneshvar E., Kousha M., Sohrabi M.S., Khataee A. and Converti A. 2012. Biosorption of three acid dyes by the brown macroalga *Stoechospermum marginatum*: Isotherm, kinetic and thermodynamic studies. *Chem. Eng. J.* 195–196: 297–306.

Davis T.A., Volesky B. and Mucci A. 2003. A review of the biochemistry of heavy metal biosorption by brown algae. *Water Res.* 37: 4311–4330.

Day R., Denizli A. and Arica M.Y. 2001. Biosorption of cadmium(II), lead(II) and copper(II) with the filamentous fungus *Phanerochaete chrysosporium*. *Bioresour. Technol.* 76: 67–70.

de Vargas I., Macaskie L.E. and Guibal E. 2004. Biosorption of palladium and platinum by sulfate reducing bacteria. *J. Chem. Technol. Biotechnol.* 79: 49–56.

Deepaa K., Chandran P. and Khan S.S. 2013. Bioremoval of direct red from aqueous solution by *Pseudomonas putida* and its adsorption isotherms and kinetics. *Ecol. Eng.* 58: 207–213.

Deniz F. and Saygideger S.D. 2010. Equilibrium, kinetic and thermodynamic studies of acid orange 52 dye biosorption by *Paulownia tomentosa* Steud. Leaf powder as a low-cost natural biosorbent. *Bioresour. Technol.* 101: 5137–5143.

Dhir B., Nasim S.A., Sharmila P. and Saradhi P.P. 2010. Heavy metal removal potential of dried *Salvinia* biomass. *Intern. J. Phytoremed.* 12: 133–141.

Elangovan R., Philip L. and Chandraraj K. 2008. Biosorption of hexavalent and trivalent chromium by palm flower (*Borassus aethiopum*). *Chem. Eng. J.* 141: 99–111.

El-All A.B.D., Azza A.M., Elham M.A. and Hanan A.M.H. 2011. Bioaccumulation of heavy metals by the water fern *Azolla pinnata*. *Egy. J. Agr. Res.* 89: 1261–1275.

Eram S., Shabbir R., Zafar H. and Javaid M. 2015. Biosorption and bioaccumulation of copper and lead by heavy metal-resistant fungal isolates. *Arab. J. Sci. Eng.* 40: 1867–1873.

Freitas H., Prasad M.N.V. and Pratas J. 2004. Plant community tolerant to trace elements growing on the degraded soils of Sao Domingos mine in the south east of Portugal: Environmental implications. *Environ. Int.* 30: 65–72.

Friedman B.A. and Dugan P.R. 1968. Concentration and accumulation of metallic ions by the bacterium *Zoogloea*. *Dev. Ind. Microbiol.* 9: 381–388.

Gadd G.M. 2009. Biosorption: Critical review of scientific rationale, environmental importance and significance for pollution treatment. *J. Chem. Technol. Biotechnol.* 84: 13–28.

Garcia-Rosales G., Olguin M.T., Colin-Cruz A. and Romero-Guzman E.T. 2012. Effect of the pH and temperature on the biosorption of lead (II) and cadmium (II) by sodium-modified stalk sponge of *Zea mays*. *Environ. Sci. Poll. Res.* 19: 177–185.

Gardea-Torresdey J.L., Tiemann K.J., Armendariz V., Bess-Oberto L., Chianelli R.R., Rios J., Parsons J.G. and Gamez G. 2000. Characterization of chromium (VI) binding and reduction to chromium (III) by the agricultural byproduct of *Avena monida* (oat) biomass. *J. Hazard. Mater.* 80: 175–188.

Mashkani S.G. and Ghazvini P.T.M. 2009. Biotechnological potential of *Azolla filiculoides* for biosorption of Cs and Sr: Application of micro-PIXE for measurement of biosorption. *Bioresour. Technol.* 100: 1915–1921.

Goodman G.T. and Roberts T.M. 1971. Plants and soils as indicators of metals in the air. *Nature* 231: 287–292.

Gould M.S. and Genetelli E.J. 1984. Effects of competition on heavy metal binding by anaerobically digested sludges. *Water Res.* 18: 123–126.

Gratao P.L., Prasad M.N.V., Cardoso P.F., Lea P.J. and Azevedo R.A. 2005. Phytoremediation: Green technology for the cleanup of toxic metals in the environment. *Braz. J. Plant Physiol.* 17: 53–64.

Gürel L. 2017. Biosorption of textile dye reactive blue 221 by capia pepper (*Capsicum annuum* L.) seeds. *Water Sci. Technol.* 75: 1889–1898.

Gutiérrez C., Hansen H.K., Hernández P. and Pinilla C. 2015. Biosorption of cadmium with brown macroalgae. *Chemosphere* 138: 164–169.

Hanif M.A., Nadeem R., Bhatti H.N., Ahmad N.R. and Ansari T.M. 2007. Ni (II) biosorption by *Cassia fistula* (Golden Shower) biomass. *J. Hazard. Mater.* 139: 345–355.

He J. and Chen J.P. 2014. A comprehensive review on biosorption of heavy metals by algal biomass: Materials, performances, chemistry, and modeling simulation tools. *Bioresour. Technol.* 160: 67–78.

Ho Y.S., Wase D.A. and Forster C.F. 1995. Batch nickel removal from aqueous solution by sphagnum moss peat. *Water Res.* 29: 1327–1332.

Horsfall M. Jr., Abia A.A. and Spiff A.I. 2006. Kinetic studies on the adsorption of Cd^{2+}, Cu^{2+} and Zn^{2+} ions from aqueous solutions by cassava (*Manihot esculenta* Cranz) tuber bark waste. *Bioresour. Technol.* 97: 283–291.

Huber A.L., Holbein B.E. and Kidby D.K. 1990. Metal uptake by synthetic and biosynthetic chemicals. In: Volesky B. (Ed.). *Biosorption of Heavy Metals*. CRC press, Boca Raton, FL. pp. 249–292.

Incharoensakdi A. and Kitjaharn P. 2002. Zinc biosorption from aqueous solution by a halotolerant cyanobacterium *Aphanothece halophytica*. *Curr. Microbiol.* 45: 261–264.

Ji G.L., Wang J.H. and Zhang X.N. 2000. Environmental problems in soil and groundwater induced by acid rain and management strategies in China. In: Huang P.M. and Iskandar I.K. (Eds.). *Soils and Groundwater Pollution and Remediation*. CRC Press, London, UK. pp. 201–224.

Ju K., Igarashi K., Miyashita S., Mitsuhashi H., Inagaki K., Fujii S., Sawada H., Kuwubara T. and Minoda A. 2016. Effective and selective recovery of gold and palladium ions from metal wastewater using a sulfothermophilic red alga, *Galdieria sulphuraria*. *Bioresour. Technol.* 211: 759–764.

Kapoor A. and Viraraghavan T. 1995. Fungal biosorption-an alternative treatment option for heavy metal bearing wastewaters: A review. *Bioresour. Technol.* 53: 195–206.

Karthikeyan T., Rajgopal S. and Miranda L.R. 2005. Cr (VI) adsorption from aqueous solution by *Hevea brasilinesis* saw dust activated carbon. *J. Hazard. Mater.* 124: 192–199.

King P., Srinivas P., Kumar Y.P. and Prasad V.S.R.K. 2006. Sorption of copper (II) ion from aqueous solution by *Tectona grandis* Lf (teak leaves powder). *J. Hazard. Mater.* 136: 560–566.

King P., Anuradha K., Beena Lahari S., Prasanna Kumar Y. and Prasad V.S.R.K. 2007. Biosorption of zinc from aqueous solution using *Azadirachta indica* bark: Equilibrium and kinetics studies. *J. Hazard. Mater.* 152: 324–329.

Kratchovil D. and Volesky B. 1998. Advances in the biosorption of heavy metals. *Trends in Biotech.* 16: 291–300.

Kuyucak N. and Volesky B. 1988. Biosorbents for recovery of metals from industrial solutions. *Biotechnol. Lett.* 10: 137–142.

Lakshmanraj L., Gurusamy A., Gobinath M.B. and Chandramohan R. 2009. Studies on the biosorption of hexavalent chromium from aqueous solutions by using boiled mucilaginous seeds of *Ocimum americanum*. *J. Hazard. Mater.* 169: 1141–1145.

Lee K.Y., Kim K.W., Baek Y.J., Chung D.Y., Lee E.H., Lee S.Y. and Moon J.K. 2014. Biosorption of uranium (VI) from aqueous solution by biomass of brown algae *Laminaria japonica*. *Water Sci. Technol.* 70: 136–143.

Liping D., Yingying S., Hua W., Xinting S. and Xiaobin Z. 2007. Sorption and desorption of lead (II) from wastewater by green algae *Cladophora fascicularis*. *J. Hazard. Mater.* 1: 220–225.

Liu H.L., Chen B.Y., Lan Y.W. and Cheng Y.C. 2004. Biosorption of Zn (II) and Cu (II) by the indigenous *Thiobacillus thiooxidans*. *Chem. Eng. J.* 97: 195–201.

Liu M., Dong F., Kang W., Sun S., Wei H., Zhang W., Nie X., Guo Y., Huang T. and Liu Y. 2014. Biosorption of strontium from simulated nuclear wastewater by *Scenedesmus spinosus* under culture conditions: Adsorption and bioaccumulation processes and models. *Intern. J. Environ. Res. Public Health* 11: 6099–6118.

Lobo Soniya M. and Gulimane K. 2015. Biosorption of heavy metals from aqueous solution using mangrove fern *Acrostichum aureum* L. leaf biomass as a sorbent. *Intern. Res. J. Environ. Sci.* 4: 25–31.

Loukidou M.X., Karapantsios T.D., Zouboulis A.I. and Matis K.A. 2004. Diffusion kinetic study of cadmiurn (II) biosorption by *Aeromonas caviae*. *J. Chem. Technol. Biotechnol.* 79: 711–719.

Lu W.B., Shi J.J., Wang C.H. and Chang J.S. 2006. Biosorption of lead, copper and cadmium by an indigenous isolate *Enterobacter* sp J1 possessing high heavy-metal resistance. *J. Hazard. Mater.* 134: 80–86.

Luo S.L., Chen L., Chen J., Xiao X., Xu T., Wan Y., Rao C. et al. 2011. Analysis and characterization of cultivable heavy metal-resistant bacterial endophytes isolated from Cd-hyperaccumulator *Solanum nigrum* L. and their potential use for phytoremediation. *Chemosphere* 85: 1130–1138.

Mahvi A.H., Nouri J., Omrani G.A. and Gholami F. 2007. Application of *Platanus orientalis* leaves in removal of cadmium from aqueous solution. *World Appl. Sci. J.* 2: 40–44.

Mameri N., Boudries N., Addour L., Belhocine D., Lounici H., Grib H. and Pauss A. 1999. Batch zinc biosorption by a bacterial nonliving *Streptomyces rimosus* biomass. *Water Res.* 33: 1347–1354.

Martins R.J. and Boaventura R.R. 2002. Uptake and release of zinc by aquatic bryophytes (*Fontinalis antipyretica* L. ex. Hedw.). *Water Res.* 36: 5005–5012.

Martins R.J. and Boaventura R.A. 2011. Modelling of lead removal by an aquatic moss. *Water Sci. Technol.* 63: 136–142.

Maurya N.S., Mittal A.K., Cornel P. and Rother E. 2006. Biosorption of dyes using dead macro fungi: Effect of dye structure, ionic strength and pH. *Biores. Technol.* 97: 512–521.

Melgar M.J., Alonso J. and Garcia M.A. 2007. Removal of toxic metals from aqueous solutions by fungal biomass of *Agaricus Macrospores*. *Sci. Total Environ.* 385: 12–19.

Michalak I., Chojnacka K. and Witek-Krowiak A. 2013. State of the art for the biosorption process—A review. *Appl. Biochem. Biotechnol.* 170: 1389–1416.

Mishra V.K. and Tripathi B.D. 2009. Accumulation of chromium and zinc from aqueous solutions using water hyacinth (*Eichhornia crassipes*). *J. Hazard. Mater.* 164: 1059–1063.

Muthukumaran K. and Sophie Beulah S. 2010. SEM and FT-IR studies on nature of adsorption of mercury (II) and chromium (VI) from wastewater using chemically activated *Syzygium jambolanum* nut carbon. *Asian J. Chem.* 22: 7857–7864.

Nakajima A. and Tsuruta T. 2004. Competitive biosorption of thorium and uranium by *Micrococcus luteus*. *J. Radioanal. Nucl. Chem.* 260: 13–18.

Nakajima A., Yasuda M., Yokoyama H., Ohya-Nishiguchi H. and Kamada H. 2001. Copper biosorption by chemically treated *Micrococcus luteus* cells. *World J. Microbiol. Biotechnol.* 17: 343–347.

Nakajima A., Horikoshi T. and Sakaguchi T. 1982. Studies on the accumulation of heavy metal elements in biological systems. *J. Appl. Microbiol.* 16: 88–91.

Nasseri S., Mazaheri A.M., Noori S.M., Rostami K.H., Shariat M. and Nadafi K. 2002. Chromium removal from tanning effluent using biomass of *Aspergillus oryzae*. *Pak. J. Biol. Sci.* 5: 1056–1059.

Ncibi M.C., Mahjou B. and Seffen M. 2007. Kinetic and equilibrium studies of methylene blue biosorption by *Posidonia oceanica* (L.) fibres. *J. Hazard. Mater.* B139: 280–285.

Neufeld R.D. and Hermann E.R. 1975. Heavy metal removal by acclimated activated sludge. *J. Water Pollut. Control Fed.* 47: 310–329.

Nourbakhsh M., Sag Y., Ozer D., Aksu Z., Kutsal T. and Caglar A. 1994. A comparative study of various biosorbents for removal of chromium (VI) ions from industrial waste waters. *Process Biochem.* 29: 1–5.

Ozdemir G., Ozturk T., Ceyhan N., Isler R. and Cosar T. 2003. Heavy metal biosorption by biomass of *Ochrobactrum anthropi* producing exopolysaccharide in activated sludge. *Bioresour. Technol.* 90: 71–74.

Ozturk A. 2007. Removal of nickel from aqueous solution by the bacterium *Bacillus thuringiensis*. *J. Hazard. Mater.* 147: 518–523.

Panda G.C., Das S.K., Chatterjee S., Maity P.B., Bandopadhyay T.S. and Guha A.K. 2006. Adsorption of cadmium on husk of *Lathyrus sativus*: Physico-chemical study. *Colloids Surf B: Biointerfaces* 50: 49–54.

Park D., Yun Y-S. and Park J.M. 2010. The past, present and future trends of biosorption. *Biotechnol. Bioprocess Eng.* 15: 86–102.

Pavasant P., Apiratikul R., Sungkhum V., Suthiparinyanont P., Wattanachira S. and Marhaba T.F. 2006. Biosorption of Cu^{2+}, Cd^{2+}, Pb^{2+}, and Zn^{2+} using dried marine green macroalga *Caulerpa lentillifera*. *Bioresour. Technol.* 97: 2321–2329.

Rajkumar M., Prasad M.N.V., Freitas H. and Ae N. 2009. Biotechnological applications of serpentine bacteria for phytoremediation of heavy metals. *Crit. Rev. Biotech.* 29: 120–130.

Rani P.S., Priya R.L. and Velan M. 2013. Sorption behavior of freshwater aquatic fern *Azolla filiculoides* on redwine dye. *Desalin. Water Treat.* 51: 31–33.

Ranjusha V.P., Pundir R., Kumar K., Dastidar M.G. and Sreekrishnan T.R. 2010. Biosorption of Remazol Black B dye (Azo dye) by the growing *Aspergillus flavus. J. Environ. Sci. Health A* 45: 1256–1263.

Rehman R., Anwar J. and Mahmud T. 2013. Sorptive removal of lead (II) from water using chemically modified mulch of *Madhuca longifolia* and *Polyalthia longifolia* as novel biosorbents. *Desalin. Water Treat.* 51: 2624–2634.

Romera E., Gonzalez F., Ballester A., Blazquez M.L. and Munoz J.A. 2006. Biosorption with algae: A statistical review. *Crit. Rev. Biotechnol.* 26: 223–235.

Romero M.C., Reinoso E.H., Urrutia M.I. and Kiernan A.M. 2006. Biosorption of heavy metals by *Talaromyces helicus*: A trained fungus for copper and biphenyl detoxification. *Electron. J. Biotechnol.* 9: 221–226.

Ruchoft C.C. 1949. The possibilities of disposal of radioactive wastes by biological treatment methods. *Sewage Works J.* 21: 877–883.

Saeed A. and Iqbal M. 2003. Bioremoval of Cd from aqueous solution by black gram husk (*Cicer arientinum*). *Water Res.* 37: 3472–3480.

Sakaguchi T., Nakajima A. and Horikoshi T. 1978. Studies on the accumulation of heavy metal elements in biological systems: VI. Uptake of uranium from sea water by microalgae. *J. Ferment. Technol.* 56: 561–565.

Salehizadeh H. and Shojaosadati S.A. 2003. Removal of metal ions from aqueous solution by polysaccharide produced from *Bacillus firmus. Water Res.* 37: 4231–4235.

Sánchez-Galván G. and Ramírez-Núñez P.A. 2014. Cationic dye biosorption by *Salvinia minima*: Equilibrium and kinetics. *Water Air Soil Poll.* 225: 2008.

Sari A., Mendil D., Tuzen M. and Soylak M. 2009. Biosorption of palladium (II) from aqueous solution by moss (*Racomitrium lanuginosum*) biomass: Equilibrium, kinetic and thermodynamic studies. *J. Hazard. Mater.* 162: 874–879.

Sari A. and Tuzen M. 2009. Removal of mercury (II) from aqueous solution using moss (*Drepanocladus revolvens*) biomass: Equilibrium, thermodynamic and kinetic studies. *J. Hazard. Mater.* 171: 500–507.

Say R., Yilmaz N. and Denizli A. 2004. Removal of chromium (VI) ions from synthetic solutions by the fungus *Penicillium purpurogenum. Eng. Life Sci.* 4: 276–280.

Selatnia A., Boukazoula A., Kechid N., Bakhti M.Z., Chergui A. and Kerchich Y. 2004. Biosorption of lead (II) from aqueous solution by a bacterial dead *Streptomyces rimosus* biomass. *Biochem. Eng. J.* 19: 127–135.

Strandberg G.W., Shumate II S.E. and Parrot J.R. 1981. Microbial cells as biosorbents of heavy metals: Accumulation of uranium by *Saccharomyces cerevisiae* and *Pseudomonas aeruginosa. Appl. Environ. Microbiol.* 41: 237–245.

Sud D., Mahajan G. and Kaur M.P. 2008. Agricultural waste material as potential adsorbent for sequestering heavy metal ions from aqueous solutions—A review. *Bioresour. Technol.* 99: 6017–6027.

Suyamboo B.K. and Srikrishnaperumal R. 2014. Biosorption of crystal violet onto *Cyperus rotundus* in batch system: Kinetic and equilibrium modeling. *Desalin. Water Treat.* 52: 4492–4507.

Tangaromsuk J., Pokethitiyook P., Kruatrachue M. and Upatham E.S. 2002. Cadmium biosorption by *Sphingomonas paucimobilis* biomass. *Bioresour. Technol.* 85: 103–105.

Tsezos M. and Volesky B. 1981. Biosorption of uranium and thorium. *Biotechnol. Bioeng.* 23: 583–604.

Umali L.J., Duncan J.R. and Burgess J.E. 2006. Performance of dead *Azolla filiculoides* biomass in biosorption of Au from wastewater. *Biotechnol. Lett.* 28: 45–50.

Uslu G. and Tanyol M. 2006. Equilibrium and thermodynamic parameters of single and binary mixture biosorption of lead (II) and copper (II) ions onto *Pseudomonas putida*: Effect of temperature. *J. Hazard. Mater.* 135: 87–93.

Vegliö F. and Beolchini F. 1997. Removal of metals by biosorption: A review. *Hydrometallurgy* 44: 301–316.

Vieira R.H.S.F. and Volesky B. 2000. Biosorption: A solution to pollution? *Intern. Microbiol.* 3: 17–24.

Vijayaraghavan K. and Yun Y.S. 2007. Chemical modification and immobilization of *Corynebacterium glutamicum* for biosorption of reactive black 5 from aqueous solution. *Ind. Eng. Chem. Res.* 46: 608–617.

Vijayaraghavan K. and Yun Y.S. 2008. Bacterial biosorbents and biosorption. *Biotechnol. Adv.* 26: 266–291.

Vijayaraghavan K. and Balasubramanian R. 2015. Is biosorption suitable for decontamination of metal-bearing wastewaters? A critical review on the state-of-the-art of biosorption processes and future directions. *J. Environ Manag.* 160: 283–296.

Vijayaraghavan K., Jegan J., Palanivelu K. and Velan M. 2005. Biosorption of copper, cobalt and nickel by marine green alga *Ulva reticulate* in a packed column. *Chemosphere* 60: 419–426.

Volesky B. and Holan Z.R. 1995. Biosorption of heavy metals. *Biotechnol. Prog.* 11: 235–250.

Volesky B., May H. and Holan Z.R. 1993. Cadmium biosorption by *Saccharomyces cerevisiae*. *Biotechnol. Bioeng.* 41: 826–829.

Volesky B. and Naja G. 2005. Biosorption: Application strategies. *16th Intern. Biotechnol. Symp. Compress Co.*, Cape Town, South Africa.

Volesky B. 2007. Biosorption and me. *Water Res.* 41: 4017–4029.

Volesky B. and Tsezos M. 1982. Separation of uranium by biosorption. US Patent US04320093.

Wang J. and Chen C. 2009. Biosorbents for heavy metals removal and their future. *Biotech. Adv.* 27: 195–226.

Wong M.H. 2003. Ecological restoration of mine degraded soils, with emphasis on metal contaminated soils. *Chemosphere* 50: 775–780.

Xiao-Ming L., De-Xiang L., Xuen-Qin X., Qi Y., Guang-Ming Z., Wei Z. and Liang G. 2008. Kinetic studies for the biosorption of lead and cooper ions by *Penicillium simplecissimun* immobilized within loofa sponge. *J. Hazard. Mater.* 159: 610–615.

Yan G. and Viraraghavan T. 2008. Mechanism of biosorption of heavy metals by *Mucor rouxii*. *Eng. Life Sci.* 8: 363–371.

Yang J., Cao J., Xing G. and Yuan H. 2014. Lipid production combined with biosorption and bioaccumulation of cadmium, copper, manganese and zinc by oleaginous microalgae *Chlorella minutissima* UTEX2341. *Bioresour Technol.* 175C: 537–544.

Zhou M., Liu Y.G., Zeng G.M., Li X., Xu W.H. and Fan T. 2007. Kinetic and equilibrium studies of Cr (VI) biosorption by dead *Bacillus licheniformis* biomass. *World J. Microbiol. Biotechnol.* 23: 43–48.

2 Quorum Sensing
Novel Bacterial Communication

Shilpi Srivastava

CONTENTS

2.1 THE MICROBIAL WORLD

Microorganisms are ubiquitous microscopic entities that inhabit any environment that is suitable for higher life forms and inhospitable settings that the majority of higher life forms would find unsuitable for survival. The assemblage includes microscopic algae, bacteria, fungi, viruses, and protozoa, all of which are too small to be seen with the naked eye. Microorganisms have the unique ability to tolerate inhospitable conditions such as temperature extremes, pressure, drought, acidity, and salinity, with some of them often requiring these conditions for survival. The metabolic versatility, phenotypic plasticity, and the unique ability to position them in a niche enable the microorganisms to persist and thrive throughout the biosphere.

The complex microbial communities are known to play an important role in maintaining the biogeochemical cycles through the production and degradation of organic matter, degradation of environmental pollutants, and recycling of nutrients. Numerous microbes are known to form an important component of the food chain in oceans, lakes, rivers, and terrestrial environments. The microorganisms help in recycling of elements by playing a key role in the breakdown of wastes and detritus and incorporate nitrogen gas from the atmosphere into organic compounds. Microbes also form a part of the alimentary canal of many animals, including man, and aid in digestion and synthesis of some vitamins (vitamins B and K). Many products made with the help of microbes such as beverages, vinegar, cheese, bread, yoghurt, buttermilk, and pickles are used by the food industry. Microbes are also commercially important since they are used in the synthesis of several industrial products such as alcohols, organic acids, acetone, enzymes, and pharmaceutical compounds.

2.2 BIOFILMS

Microorganisms rapidly adapt to changing environmental parameters by exploiting intercellular interactions and various communication mechanisms. This complex differentiation and collective approach has been reported for a number of different microbes under diverse environmental conditions. Most microorganisms are primarily characterized as planktonic, freely suspended cells.

A biofilm is a sessile community of single or multiple microbial components cocooned in a self-produced matrix of extracellular polymeric substances (EPS) that allows the microbe to flourish in sessile environments. Biofilms can be formed by single or multiple microbial components associated with a surface and enclosed in a matrix that allows the growth and survival in sessile environments. Mixed-species biofilms are found in most environments, whereas single-species biofilms exist in a variety of infections and on the surface of medical implants. Apart from living forms, a number of noncellular materials such as mineral crystals, soil particles (clay or silt), blood components, and particles produced by corrosion may also be found in the biofilm matrix. Recent researches have shown that other microorganisms such as protozoa, fungi, and microalgae may also be a core component of biofilms (Figure 2.1). The formation of biofilms has been an ancient and integral part of prokaryotic development. During the evolutionary history, biofilms provided protection and balance against extreme environmental conditions, which facilitated the development of complex interactions between individual cells and provided an environment that was beneficial for the development of signaling pathways and chemotactic motility.

Anton Van Leeuwenhoek is credited with the discovery of microbial biofilms while observing microorganisms on tooth surfaces by using his simple microscope in 1684. Leeuwenhoek termed the organisms in dental plaque *animalcules*.

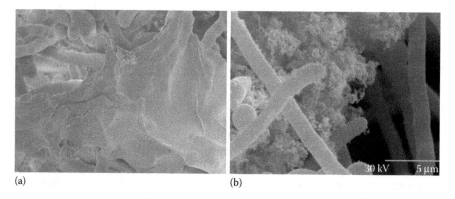

(a)   (b)

FIGURE 2.1 A cryo-scanning electron micrograph (SEM) of an *Aspergillus fumigatus* biofilm (a) and an SEM view of *Candida albicans in vitro* biofilm cells (b). (Reprinted from Fanning, S. and Mitchell, A.P., *PLoS Pathol.*, 8, e1002585, 2012.)

About 200 years later, Hauser (1885) observed that swarming allowed the bacteria to effectively colonize the entire surface of the agar plate. The *bottle effect* was observed by Heukelekian and Heller (1940) while studying marine microorganisms, and the research was published in the *Journal of Bacteriology*. Bottle effect refers to an increase in the bacterial activity and growth in the presence of a surface to which the bacteria can attach. Around the same time, Zobell (1943) observed that the bacteria introduced in glass bottles disappeared from the liquid phase, along with an increase in the number of bacteria attached to the surface of the bottles. ZoBell also noticed that attachment was a two-phase process that consisted of a primary phase and a reversible phase of attraction to the surface, followed by the second phase of irreversible adhesion. Research on biofilms accelerated after the advent of advanced tools such as scanning electron microscope (SEM) and transmission electron microscope (TEM), which resulted in high-resolution photomicroscopy at much higher magnifications. Finally, in 1978, Costerton and colleagues published their monumental work entitled "How Bacteria Stick," which revolutionized the research on biofilms. Costerton and his research group formulated a theory of biofilms that explained the underlying mechanism of why microorganisms adhered to living and nonliving materials. Therefore, although biofilm was initially defined as the formation of a community of microorganisms attached to a surface, Costerton's work modified the concept of biofilms as a complex developmental process that is multifaceted and dynamic in nature. Different approaches such as staining, metabolic, genetic, mass spectrometry, and advances microscopic techniques that aid in the visualization and quantification of biofilms (Figure 2.2) are now available. Real-time quantitative reverse transcription polymerase chain reaction (qRT-PCR) and fluorescence *in situ* hybridization (FISH) are the most recently developed techniques for the detection and quantification of microorganisms in biofilms.

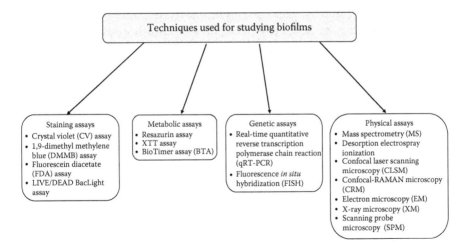

FIGURE 2.2 Different techniques used for detailed investigation of biofilms. (With kind permission from Springer Science+Business Media: *Biotechnol. Lett.*, Biofilms and human health, 38, 2016, 1–22, Srivastava, S. and Bhargava, A.)

2.3 QUORUM SENSING

Cooperation is at the core of social behavior, where all individuals contribute and gain more or less equally. Cooperation can also involve division of labor, whereby individuals engage in different tasks from which they might obtain different benefits directly or via benefits to kin. Cooperation and communication are the key to success in microorganisms, wherein they come together to perform a wide range of functions such as dispersal, foraging, chemical warfare, biofilm formation, and quorum sensing (QS), which are normally seen in multicellular organisms. Two essential conditions must be met for true communication to occur. First, a signal must be produced by one or more individuals that can be perceived by other individuals, and second, the perceivers must respond by alteration in their behavior. Communication in bacteria usually takes place through the following two mechanisms:

1. Signals produced at different stages of growth by the bacteria, which are not constrained by cell density requirements.
2. Signals related to cell density and populations, usually referred to as *quorum-sensing* signals. These are utilized by bacteria in a concentration-dependent manner.

Quorum sensing is a process of bacterial cell–cell communication, which involves the production, detection, and response of the microbe to extracellular signaling molecules known as autoinducers (AIs) (Figure 2.3). Quorum sensing was first described in the marine bioluminescent bacterium *Vibrio fischeri*, a heterotrophic gram-negative, rod-shaped bacterium found living in the squids. The bacteria live in the light organs of these marine animals and gain nutrients and shelter from their hosts. In turn, the microbes produce luminescence, which helps the animals escape from predators. In 1970, K.H. Nealson and J.W. Hastings observed that the bacterial cells would not light up until they reached a certain threshold population density during the late logarithmic phase of growth, which avoided wasting of their resources in producing light when illumination would be ineffective.

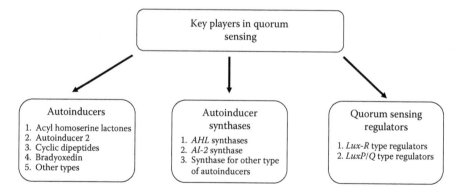

FIGURE 2.3 Different components involved in quorum sensing.

It was concluded that bioluminescence in these bacteria was probably regulated by chemical messengers that are transferred between cells. A signal that indicates sufficient cells in the vicinity triggers a response cascade, which alters the expression of numerous genes, some of which have no known function. When the bacteria have low population densities, the *AI* synthase gene expresses at the base level and results in the synthesis of minute quantity of AI signal molecules that diffuse out of the cell but are diluted in the surrounding environment in a short period of time. However, an increase in the bacterial population results in gradual accumulation of AIs in and around the bacterial cells. These AI molecules activate a transcriptional regulator protein by binding to it, which then interacts with target DNA sequences and enhances or blocks the transcription of QS-regulated genes. This results in the synchronous activation of certain phenotypes in a bacterial population. By the 1980s, autoinduction was proved in *V. fischeri* at the physiological, chemical, and genetic levels. Quorum sensing is not a new phenomenon but is of ancient origin, with many components of the QS systems typically encoded by nuclear genes that were acquired by the microbe through horizontal gene transfer. The QS systems integrate with the native signal-transduction systems after successful transfer to the new microbial genome and thereafter produce regulatory networks, which are often unique to a given species. Sifri (2008) has divided QS into the following four steps:

1. Small biochemical signal molecules called AIs are produced by the bacteria
2. Release of these AIs into the surrounding environment, either actively or passively
3. Specific receptors recognize the signal molecules once their concentration exceeds a threshold level
4. Changes in gene regulation

Both gram-negative and gram-positive bacteria use QS communication for regulating a wide range of physiological activities that include attachment to the surface, formation of spores or fruiting bodies, bioluminescence, synthesis of antibiotics, storage of nutrients, biosurfactant synthesis, extracellular polymer production, competence, programmed cell death, swimming and twitching motility, and the secretion of nutrient-sequestering compounds (Table 2.1).

2.4 AUTOINDUCERS

The phenomenon of QS is dependent on the interaction of small diffusible signal molecules called AIs with a sensor or transcriptional activator. This initiates gene expression, resulting in coordinated activities. Quorum sensing is initiated by the release and detection of AI molecules, whose concentration increases with increasing cell population density. The AIs are produced at low levels in the beginning, but their concentration increases with growth. The microbes sense this information to induce changes in their cell numbers and thereafter alter their activity and gene expression. Processes that are unproductive and costly when undertaken by a single bacterial cell are usually controlled by QS to make them resource-effective for

TABLE 2.1

Quorum Sensing Systems in Different Bacteria and Their Role in Biofilm Development

Bacteria	+/−	Quorum-Sensing System	Role in Biofilm Development	Reference
Salmonella enterica	−	LuxS	Attachment	Prouty et al. (2002)
Serratia liquefaciens	−	Acyl-HSL	Maturation	Labbate et al. (2004)
Burkholderia cepacia	−	cepI/R	Maturation	Huber et al. (2001)
Aeromonas hydrophila	−	ahyR/I acyl-HSL	Maturation	Lynch et al. (2002)
Pseudomonas aeruginosa	−	LuxI/LuxR; PQS	Maturation	Holm and Vikström (2014)
Pseudomonas aureofaciens	−	PhzI/PhzR	Phenazine antibiotic	Pierson et al. (1994)
Streptococcus mutans	+	LuxS	Maturation	Merritt et al. (2003) Wen and Burne (2004)
Rhodobacter sphaeroides	−	Acyl-HSL	Cellular aggregation	Puskas et al. (1997)
Yersinia pseudotuberculosis	−	acyl-HSL ypsI/R	Maturation	Atkinson et al. (1999)
Xanthomonas campestris	−	DSF/rpf	Dispersal	Slater et al. (2000) Dow et al. (2003)
Vibrio fischeri	−	LuxI/LuxR	Bioluminescence	Engebrecht and Silverman (1987)
Vibrio harveyi	−	LuxLM/LuxN	Bioluminescence	Bassler et al. (1994)
Agrobacterium tumefaciens	−	TraI/TraR	Ti plasmid conjugation	Piper et al. (1993); Zhang et al. (1993)
Escherichia coli	−	LuxR/SdiA	Cell division, attachment, and effacing lesion formation	Withers and Nordstrom (1998) Sperandio et al. (1999)

the microbe. Autoinducers are low-molecular-weight molecules and are synthesized intracellularly. These molecules are either passively released or actively secreted outside of the cells. Increase in the number of microbial cells in a population concomitantly increases the extracellular concentration of the AI. The accumulation of AIs above the minimal threshold level required for detection triggers a signal transduction cascade, resulting in population-wide changes in gene expression.

Autoinducers and other such molecules that are produced by both eukaryotic and prokaryotic organisms could be used for one-, two-, or multiway communication. Apart from their importance in QS in prokaryotes, AIs provide probiotic functions, alter the composition of the microbiota, affect the expression of virulence genes, and encourage pathogens to disperse from biofilms in eukaryotic hosts.

Although the AI activity is specific, several bacteria are known to produce and detect several AIs. The bioluminescent marine bacterium *Vibrio harveyi* uses three AIs to regulate approximately 600 target genes at intraspecific, intrageneric, and interspecies levels. *Pseudomonas aeruginosa* uses two canonical acyl-homoserine lactone (AHL) AIs and non-AHL AIs for QS.

2.5 QUORUM SENSING IN GRAM-NEGATIVE BACTERIA

Ng and Bassler (2009) have discussed four common features found in almost all known gram-negative QS systems.

1. The AIs are AHLs or other molecules that are synthesized from *S*-adenosyl methionine (SAM), and thereafter, they diffuse readily through the bacterial membrane.
2. Autoinducers are bound by specific receptors that reside either in the inner membrane or in the cytoplasm.
3. Quorum sensing typically alters dozens to hundreds of genes that underpin various biological processes.
4. There is an increased synthesis of the AI through an AI-driven activation of QS, known as autoinduction. This establishes a feed-forward loop that possibly promotes synchronous gene expression in the population.

Acyl-homoserine lactones, having a core *N*-acylated homoserine-lactone ring carrying acyl chain of C_4 to C_{18} in length, with modifications at the C_3 position or unsaturated double bonds, are the most common class of AIs in gram-negative bacteria (Figure 2.4). Only in rare cases, an AI of one bacterial species influences the expression of target genes in another species. This signaling specificity is controlled by two primary factors: the substrate specificity of the *LuxI*-like proteins and the specificity in the binding of the *LuxR*-type proteins to their cognate AHLs.

In most cases, the *LuxI* enzymes produce AHLs by deriving the lactone moiety from SAM, with the specific acyl chain obtained from fatty acid anabolism. However, some exceptions are also found. In *Rhodopseudomonas palustris*, the photosynthetic bacterium, 4-coumaroyl-homoserine lactone synthase (*RpaI*) produces *p*-coumaroyl-homoserine lactone (HSL), for which the acyl group comes from *p*-coumarate, a host metabolite. Some plant-associated bacteria synthesize unusual HSL AIs by using bacterial substrates. For example, *Aeromonas* spp. and *Bradyrhizobium japonicum* produce isovaleryl-HSL, and *Bradyrhizobium* BTAi produces cinnamoyl-HSL. Similarly, atypical AIs have also been reported in some microbes such as *Ralstonia solanacearum* and *Xanthomonas campestris*. The *PhcB* protein of *R. solanacearum* synthesizes one of the two related AIs that control virulence and the formation of biofilms, viz. 3-hydroxypalmitic-acid-methyl-ester (3-OH PAME) and (*R*)-methyl-3-hydroxymyristate ((*R*)-3-OH MAME). *Xanthomonas campestris* uses a diffusible signal factor *cis*-11-methyl-2-dodecenoic acid to facilitate transitions between its planktonic and biofilm-associated lifestyles.

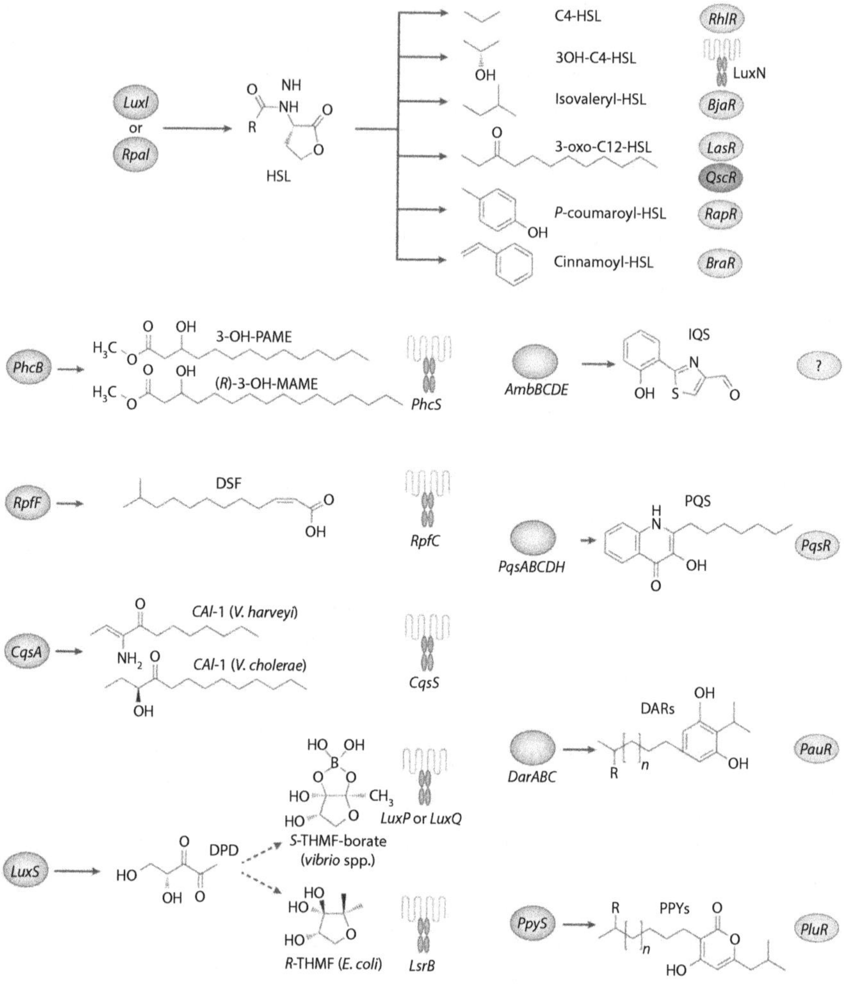

FIGURE 2.4 Quorum-sensing synthases, autoinducers, and receptors. (Reprinted by permission from Macmillan Publishers Ltd. *Nat. Rev. Microbiol.*, Papenfort, K. and Bassler, B.L., 2016, copyright 2016.)

However, several issues such as the presence of several AIs, mixed species consortia, internal changes, and external fluctuations need to be sorted out for accurate execution of QS. Systems biology has shed light on these aspects by uncovering common network design principles that occur in QS systems. Figures 2.5 and 2.6 show the QS circuits in *P. aeruginosa* and *V. harveyi*. There are currently four well-known QS pathways in *P. aeruginosa* with cytoplasmic DNA-binding transcription factors. The QS systems in *P. aeruginosa* are organized in a hierarchy, with *LasR* at the top of the cascade and receptors. *Vibrio harveyi* and *Vibrio cholerae* provide the second examples of a canonical QS circuit that relies on membrane-bound receptors.

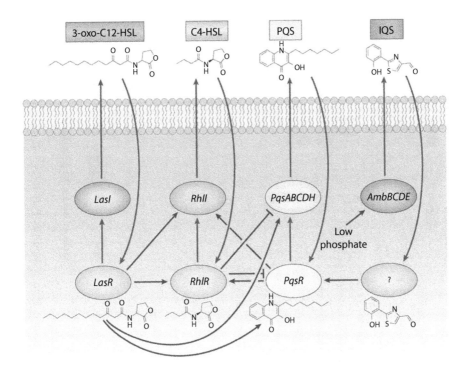

FIGURE 2.5 Quorum-sensing circuits in *Pseudomonas aeruginosa*. (Reprinted by permission from Macmillan Publishers Ltd. *Nat. Rev. Microbiol.*, Papenfort, K. and Bassler, B.L., 2016, copyright 2016.)

2.6 QUORUM SENSING IN GRAM-POSITIVE BACTERIA

Gram-positive bacteria use secreted oligopeptides and two-component systems consisting of membrane-bound sensor kinase receptors and cognate cytoplasmic response regulator, which functions as a transcriptional regulator and directs alterations in gene expression.

Two types of QS systems have been reported in gram-positive bacteria (Figure 2.7). The first system comprises three components: an autoinducing peptide (AIP) that is basically a signaling peptide and a two-component signal transduction system (TCSTS) that detects and responds to an AIP with definite specificity. Usually, a signal peptide precursor is produced by gram-positive bacteria, which is cleaved from the double-glycine consensus sequence, and the resultant active AIP is then transferred into their environments through a peptide-specific ABC transporter. Gram-positive bacteria contain signaling peptides that consist of 5–25 amino acids, of which some have unusual side chains. A TCSTS mediates the detection of signaling peptides in gram-positive bacteria. It consists of a membrane-associated, histidine kinase protein, which senses the AIP, and a cytoplasmic response regulator protein, which enables the cell to respond to the peptide via regulation of gene expression.

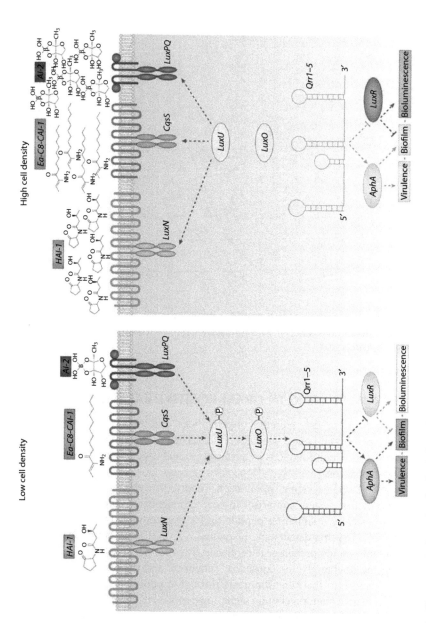

FIGURE 2.6 Quorum-sensing circuits in *Vibrio harveyi*. (Reprinted by permission from Macmillan Publishers Ltd., *Nat. Rev. Microbiol.*, Papenfort, K. and Bassler, B.L., 2016, copyright 2016.)

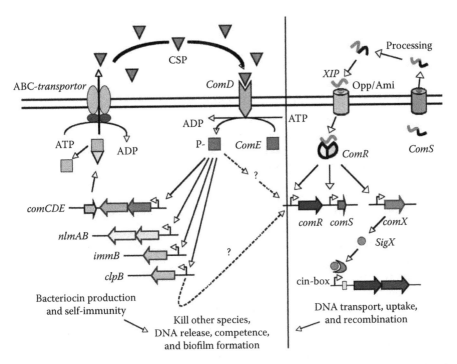

FIGURE 2.7 A schematic diagram indicating two types of signaling peptide-mediated quorum-sensing systems in gram-positive bacterium, *Streptococcus mutans*. (From Li, Y.-H. and Tian, X., *Sensors*, 12, 2519–2538, 2012.)

2.7 QUORUM SENSING AND VIRULENCE

It has become established that QS is used by many bacterial species as an important component of their regulatory machinery, with QS controlling virulence gene expression in numerous microorganisms. The QS systems are an integral part of the key virulence regulators in gram-negative and gram-positive bacteria. Quorum sensing occupies a pivotal position in bacterial pathogenicity, and therefore, its key role in the virulence of multiple human pathogens has been exhaustively studied.

> *Staphylococcus aureus*, a component of the normal human skin flora (found in about 30% of the adult human population), can cause minor skin infections, which can lead to bacteremia, pneumonia, and sepsis. Several infections of *S. aureus* such as osteomyelitis, endocarditis, and foreign-body related infections are due to the formation of biofilms rather than by free-living cells. The accessory gene regulator (*agr*) QS system has been assigned an important role in the model of *S. aureus* pathogenesis. Staphylococcal infections include many virulence factors such as toxins and autolysins, adhesion molecules, hemolysin, and compounds affecting the immune system, with QS influencing it via the *agr* system. The *agr*

system is intricately involved in the regulation of virulence genes, predominantly from two promoters, P2 and P3, which produce *RNAII* and *RNAIII*, respectively. The *agr* system is thought to regulate about 23 genes contributing to bacterial virulence. Two classes of the virulence factors are regulated by *agr*: the first is composed of virulence factors involved in evasion of the host immune system and the resulting attachment to the host, whereas the second class comprises genes involved in the production of exoproteins associated with invasion and toxin production. The activation of the *agr* system is considered a prerequisite for switching the bacterium from an adhesive, colonizing commensal to an aggressive pathogen that is invasive in nature.

Streptococcus mutans provides another example of bacteria using QS for biofilm formation and effective virulence. It normally lives in its natural ecosystem, that is, the dental plaque. However, under certain appropriate conditions, *S. mutans* utilizes dietary fermentable carbohydrates to produce acids and initiate demineralization of the tooth surface or dental caries. It is due to this that it is considered an important etiological agent of dental caries. *Streptococcus mutans* has a well-conserved QS system comprising about three gene products encoded by *ComCDE*. The most remarkable feature in *S. mutans* is that *ComCDE* QS system appears to influence the key virulence factors in the pathogenesis of this organism, viz. bacteriocin production, stress response, genetic competence, and biofilm formation.

Bacillus cereus causes both intestinal and nonintestinal infections in humans and is most commonly associated with food poisoning. A number of toxins, hemolysins, and phospholipases secreted by the microbe cause acute diarrheal diseases in human beings. The presence of the transcription factor PlcR is a prerequisite for QS in *B. cereus*, since it controls the expression of most of the virulence factors of the bacteria after binding to the intracellular AIPs derived from the *PapR* protein.

Pseudomonas aeruginosa, an opportunistic gram-negative bacterium commonly associated with nosocomial infections as well as found in the infections of severely burned individuals, is a major cause of death by severe respiratory infections. *Pseudomonas aeruginosa* infections are difficult to eradicate because the bacteria show high levels of resistance to antibiotics and their ability to form biofilms. The bacteria use several QS systems to survive the inhospitable conditions on the surface and within the host, as well as to circumvent the immune system of the host to cause disease. *Pseudomonas aeruginosa* has three QS systems: two *LuxI/LuxR*-type QS circuits that control the expression of virulence factors, along with a third, non-*LuxI/LuxR*-type system, known as the *Pseudomonas* quinolone signal (PQS). The QS-regulated virulence factors in *P. aeruginosa* include elastase, proteases, pyocyanin, lectin, swarming motility, rhamnolipids, and toxins. The accessory regulators of *P. aeruginosa* QS system, notably *Vfr*, *AlgR*, *PhoR/B*, *RpoN*, *RpoS*, *DksA*, *RelA*, *GidA*, *QscR*, *RsmY*, *RsmZ*, *RsmA*, *PrrF*, *PhrS*, *VqsM*, *VqsR*, and *RsaL*, presumably fine-tune the network, so that the virulence factor production occurs with optimal precision. It has been reported that up to 10% of the *P. aeruginosa* genome is controlled by QS.

Vibrio cholerae, a gram-negative bacterium, is the causative agent of cholera, an endemic disease in underdeveloped regions. Cholera causes profuse watery stools, which can lead to dehydration and death if not properly treated. The typical cholera symptoms are due to an enterotoxin called cholera toxin, expression of which is controlled by QS. *Vibrio cholerae* responds to two AIs by using two parallel QS circuits, viz. (S)-3-hyroxytridecan-4-one *(CAI-1)* and *AI-2*. Through these two AIs, the bacteria detect both the number of other vibrios and the total number of bacteria in the environment.

Thus, QS is of prime importance for the bacteria, since it enables the bacteria to appropriately time expression of immune response-activating products. Using QS, bacteria accumulate in a high cell density before virulence determinants are expressed and thereafter make a concerted attack and produce ample virulence factors to overwhelm the host defenses.

2.8 TARGETING QUORUM SENSING TO CONTROL BIOFILM INFECTIONS

Biofilms are of prime importance for human health, since they account for about 65% of all hospital infections in humans and are responsible for several infectious diseases associated with inert surfaces, including medical devices for internal or external use. The common biofilm-forming bacteria cause a number of diseases in human beings such as cystic fibrosis, native valve endocarditis, otitis media, periodontitis, and chronic bacterial prostatitis. In recent years, the control of QS has become the prime target for the development of new antibiofilm drugs.

A study in guinea pigs has confirmed that *yd 47*, a novel QS inhibitor, showed potent inhibition effect against otitis media and biofilm formation induced by *Streptococcus pneumoniae* on cochlear implants. Similar results have been obtained in rat models, when the *RNAIII*-inhibiting peptide suppressed staphylococcal TRAP/*agr* systems and reduced biofilm formation *in vivo*, which suggested the role of *RNAIII*-inhibiting peptide as an anti-quorum sensing (anti-QS) or/and antibiofilm agent. It has also been reported that QS inhibitor increased the susceptibilities of gram-negative and gram-positive biofilms to antibiotics *in vitro* and *in vivo*.

Less efficacy of the existing antibiotics in controlling microbial infection and formation of biofilms by pathogenic bacteria make it imperative to find alternatives to currently available synthetic drugs. The most recent approach involves the evaluation of QS inhibitors from natural sources. A number of plants have been evaluated that interfere with bacterial QS signaling pathways, resulting in reduced biofilm formation. The bioactive metabolite phytol obtained from *Piper betle*, a tropical creeper plant of family Piperaceae, was used to evaluate the anti-QS and antibiofilm efficacy of against *Serratia marcescens*. Ethyl acetate extract (PBE) of *P. betle* effectively inhibited the hydrophobicity, swarming motility, and downregulated the QS genes, which confirmed its anti-QS potential against *S. marcescens*. In another study, the *in vivo* inhibitory activity of *Murraya koenigii* essential oil (EO) on QS-controlled virulence factors of *P. aeruginosa* was investigated by using *Caenorhabditis elegans* as a model organism. *Murraya koenigii* EO significantly inhibited the pyocyanin production and staphylolytic LasA activity of the pathogen, both of which are under the control of the QS system. Musthafa et al. (2017) evaluated the anti-virulence potential of the medicinal plant extracts and their derived phytochemicals against *P. aeruginosa*. The plant extracts demonstrated an inhibitory activity over a virulence factor, pyoverdin, produced by *P. aeruginosa* ATCC 27853. A number of edible plants such as *Allium sativum*, *Zingiber officinale*, ginseng, pineapple, plantain, and sapodilla are being explored as QS inhibitors.

2.9 CONCLUSION

A major part of the bacterial genome and proteome is usually affected by quorum signaling, suggesting that QS is a phenomenon used by many pathogenic bacteria for adapting to the metabolic demands of living beings in communities as well as to modulate virulence factor production. The increased emergence of multiple-drug-resistant bacteria has further accelerated the finding of novel strategies for treating bacterial infections. Since a vast array of microbes use QS to control virulence factor production, it is imperative to consider QS as an attractive target for antimicrobial therapy. Pathogenic microbes that use QS to control virulence could potentially be rendered avirulent by blocking this cell-to-cell signaling mechanism. Since many microbial pathogens use QS to regulate virulence, strategies should be designed to block or interfere with these signaling systems that will have broad applicability for biological control of pathogens. Interference with QS represents a promising strategy for the therapeutic or prophylactic control of infection.

BIBLIOGRAPHY

Abdelinour A., Arvidson S., Bremell T., Ryden C. and Tarkowski A. 1993. The accessory gene regulator (*agr*) controls *Staphylococcus aureus* virulence in a murine arthritis model. *Infect. Immun.* 61: 3879–3885.

Adonizio A.L., Downum K., Bennett B.C. and Mathee K. 2006. Anti-quorum sensing activity of medicinal plants in southern Florida. *J. Ethnopharmacol.* 103: 427–435.

Atkinson S., Throup J.P., Stewart G.S. and Williams P. 1999. A hierarchical quorum-sensing system in *Yersinia pseudotuberculosis* is involved in the regulation of motility and clumping. *Mol. Microbiol.* 33: 1267–1277.

Bassler B.L., Wright M. and Silverman M.R. 1994. Multiple signaling systems controlling expression of luminescence in *Vibrio harveyi*: Sequence and function of genes encoding a second sensory pathway. *Mol Microbiol.* 13: 273–286.

Bhargava A., Gupta V.K., Singh A.K. and Gaur R. 2012. Microbes for heavy metal remediation. In: Gaur R., Mehrotra S. and Pandey R.R. (Eds.) *Microbial Applications*. IK International Publishing House, New Delhi, India, pp. 167–177.

Bjarnsholt T., Jensen P.O., Rasmussen T.B., Christophersen L., Calum H., Hentzer M., Hougen H.-P. et al. 2005. Garlic blocks quorum sensing and promotes rapid clearing of pulmonary *Pseudomonas aeruginosa* infections. *Microbiology* 151: 3873–3880.

Bordi C. and de Bentzmann S. 2011. Hacking into bacterial biofilms. *Ann. Intensive Care* 1: 19.

Borges A., Abreu A., Malheiro J., Saavedra M.J. and Simões M. 2013. Biofilm prevention and control by dietary phytochemicals. In: Méndez-Vilas A. (Ed.) *Microbial Pathogens and Strategies for Combating Them: Science, Technology and Education.* Microbiology Book Series Volume 1. Formatex Research Center, Badajoz, Spain, pp. 32–41.

Borges A., Abreu A.C., Dias C., José Saavedra M., Borges F. and Simões M. 2016. New perspectives on the use of phytochemicals as an emergent strategy to control bacterial infections including biofilms. *Molecules* 21: 877.

Bottone E.J. 2010. *Bacillus cereus,* a volatile human pathogen. *Clin. Microbiol. Rev.* 23: 382–398.

Brackman G., Cos P., Maes L., Nelis H.J. and Coenye T. 2011. Quorum sensing inhibitors increase the susceptibility of bacterial biofilms to antibiotics *in vitro* and *in vivo. Antimicrob. Agents Chemother.* 55: 2655–2661.

Brown S.P. and Johnstone R.A. 2001. Cooperation in the dark: Signalling and collective action in quorum-sensing bacteria. *Proc. R. Soc. B.* 268: 961–965.

Cevizci R., Düzlü M., Dündar Y., Noyanalpan N., Sultan N., Tutar H. and Bayazıt Y.A. 2015. Preliminary results of a novel quorum sensing inhibitor against pneumococcal infection and biofilm formation with special interest to otitis media and cochlear implantation. *Eur. Arch. Otorhinolaryngol.* 272: 1389–1393.

Costerton J.W. 1995. Overview of microbial biofilms. *J. Ind. Microbiol.* 15: 137–140.

Costerton J.W. 1999. Introduction to biofilm. *Intern. J. Antimicrob. Agents* 11: 217–221.

Costerton J.W., Lewandowski Z., Caldwell D., Korber D. and Lappin-Scott H.M. 1995. Microbial biofilms. *Annu. Rev. Microbiol.* 49: 711–745.

Costerton J.W., Stewart P.S. and Greenberg E.P. 1999. Bacterial biofilms: A common cause of persistent infections. *Science* 284: 1318–1322.

Costerton J.W., Cheng K.-J., Geesey G.G., Ladd T.I., Nickel J.C., Dasgupta M. and Marrie T.J. 1987. Bacterial biofilms in nature and disease. *Annu. Rev. Microbiol.* 41: 435–464.

Crespi B.J. 2001. The evolution of social behavior in microorganisms. *Trends Ecol. Evol.* 16: 178–183.

Daniels R., Vanderleyden J. and Michiels J. 2004. Quorum sensing and swarming migration in bacteria. *FEMS Microbiol. Rev.* 28: 261–289.

Das N. and Chandran P. 2011. Microbial degradation of petroleum hydrocarbon contaminants: An overview. *Biotechnol. Res. Intern.* 2011: 941810.

Davey M.E. and O'Toole G.A. 2000. Microbial biofilm: From ecology to molecular genetics. *Microbiol. Mol. Biol. Rev.* 64: 847–867.

De Kievit T.R. 2008. Quorum sensing in *Pseudomonas aeruginosa* biofilms. *Environ. Microbiol.* 11: 279–288.

De Kievit T.R. and Iglewski B.H. 2000. Bacterial quorum sensing in pathogenic relationships. *Infect. Immun.* 68: 4839–4849.

Donabedian H. 2003. Quorum sensing and its relevance to infectious diseases. *J. Infect.* 46: 207–214.

Dow J.M., Crossman L., Findlay K., He Y.-Q., Feng J.-X. and Tang J.-L. 2003. Biofilm dispersal in *Xanthomonas campestris* is controlled by cell-cell signaling and is required for full virulence to plants. *Proc. Natl. Acad. Sci. (USA)* 100: 10995–11000.

Engebrecht J. and Silverman M. 1987. Nucleotide sequence of the regulatory locus controlling expression of bacterial genes for bioluminescence. *Nucleic Acids Res.* 15: 10455–10467.

Fanning S. and Mitchell A.P. 2012. Fungal biofilms. *PLoS Pathol.* 8: e1002585.

Federle M.J. and Bassler B.L. 2003. Interspecies communication in bacteria. *J. Clin. Invest.* 112: 1291–1299.

Finch R.G., Pritchard D.I., Bycroft B.W., Williams P. and Stewart G.S. 1998. Quorum sensing: A novel target for anti-infective therapy. *J. Antimicrob. Chemother.* 42: 569–571.

Fux C.A., Costerton J.W., Stewart P.S. and Stoodley P. 2005. Survival strategies of infectious biofilms. *Trends Microbiol.* 13: 34–40.

Ganesh P.S. and Rai R.V. 2016. Inhibition of quorum-sensing-controlled virulence factors of *Pseudomonas aeruginosa* by *Murraya koenigii* essential oil: A study in a *Caenorhabditis elegans* infectious model. *J. Med. Microbiol.* 65: 1528–1535.

Ghannoum M. and O'Toole G.A. 2004. *Microbial Biofilms*. ASM Press, Washington, DC.

González J.E. and Keshavan N.D. 2006. Messing with bacterial quorum sensing. *Microbiol. Mol. Biol. Rev.* 70: 859–875.

Gotz F. 2002. *Staphylococcus* and biofilms. *Mol. Microbiol.* 43: 1367–1378.

Griffin A.S., West S.A. and Buckling A. 2004. Cooperation and competition in pathogenic bacteria. *Nature* 430: 1024–1027.

Hall-Stoodley L., Costerton J.W. and Stoodley P. 2004. Bacterial biofilms: From the natural environment to infectious diseases. *Nat. Rev. Microbiol.* 2: 95–108.

Hammer B.K. and Bassler B.L. 2003. Quorum sensing controls biofilm formation in *Vibrio cholerae*. *Mol. Microbiol.* 50: 101–104.

Harraghy N., Kerdudou S. and Herrmann M. 2007. Quorum-sensing systems in *Staphylococci* as therapeutic targets. *Anal. Bioanal. Chem.* 387: 437–444.

Hastings J.W. and Greenberg E.P. 1999. Quorum sensing: The explanation of a curious phenomenon reveals a common characteristic of bacteria. *J. Bacteriol.* 181: 2667–2668.

Holm A. and Vikström E. 2014. Quorum sensing communication between bacteria and human cells: Signals, targets, and functions. *Front. Plant Sci.* 5: 309.

Horikoshi K. and Grant W.D. 1998. *Extremophiles: Microbial Life in Extreme Environments*. Wiley-Liss, New York.

Hoyle B.D. and Costerton J.W. 1991. Bacterial resistance to antibiotics: The role of biofilms. *Prog. Drug. Res.* 37: 91–105.

Huber B., Riedel K., Hentzer M., Heydorn A., Gotschlich A., Givskov M., Molin S. and Eberl L. 2001. The cep quorum-sensing system of *Burkholderia cepacia* H111 controls biofilm formation and swarming motility. *Microbiology* 147: 2517–2528.

Hughes D.T. and Sperandio V. 2008. Inter-kingdom signalling: Communication between bacteria and their hosts. *Nat. Rev. Microbiol.* 6: 111–120.

Ji C., Wang J. and Liu T. 2015. Aeration strategy for biofilm cultivation of the microalga *Scenedesmus dimorphus*. *Biotechnol. Lett.* 37: 1953–1958.

Joint I., Tait K., Callow M.E, Callow J.A, Milton D., Williams P. and Camara M. 2002. Cell-to-cell communication across the prokaryote-eukaryote boundary. *Science* 298: 1207.

Kaiser D. and Losick R. 1993. How and why bacteria talk to each other. *Cell* 73: 873–885.

Keller L. and Surette M.G. 2006. Communication in bacteria: An ecological and evolutionary perspective. *Nat. Rev. Microbiol.* 4: 249–258.

Kleerebezem M., Quadri L.E., Kuipers O.P. and de Vos W.M. 1997. Quorum sensing by peptide pheromones and two-component signal-transduction systems in gram-positive bacteria. *Mol. Microbiol.* 24: 895–904.

Kokare C.R., Chakraborty S., Khopade A.N. and Mahadik K.R. 2009. Biofilm: Importance and applications. *Ind. J. Biotech.* 8: 159–168.

Kostakioti M., Hadjifrangiskou M. and Hultgren S.J. 2013. Bacterial biofilms: Development, dispersal, and therapeutic strategies in the dawn of the post-antibiotic era. *Cold Spring Harb. Perspect. Med.* 3: a010306.

Labbate M., Queck S.Y., Koh K.S., Rice S.A., Givskov M. and Kjelleberg S. 2004. Quorum sensing-controlled biofilm development in *Serratia liquefaciens* MG1. *J. Bacteriol.* 186: 692–698.

Lazdunski A.M., Ventre I. and Sturgis J.N. 2004. Regulatory circuits and communication in gram-negative bacteria. *Nat. Rev. Microbiol.* 2: 581–592.

Lerat E. and Moran N.A. 2004. The evolutionary history of quorum-sensing systems in bacteria. *Mol. Biol. Evol.* 21: 903–913.

Li Y.H., Tang N., Aspiras M.B., Lau P.C.Y., Lee J.H., Ellen R.P. and Cvitkovitch D.G. 2002. A quorum-sensing signaling system essential for genetic competence in *Streptococcus mutans* is involved in biofilm formation. *J. Bacteriol.* 184: 2699–2708.

Li Y.-H. and Tian X. 2012. Quorum sensing and bacterial social interactions in biofilms. *Sensors* 12: 2519–2538.

LoVetri K. and Madhyastha S. 2010. Antimicrobial and antibiofilm activity of quorum sensing peptides and peptide analogues against oral biofilm bacteria. *Methods Mol. Biol.* 618: 383–392.

Lynch M.J., Swift S., Kirke D.F., Keevil C.W., Dodd C.E.R. and Williams P. 2002. The regulation of biofilm development by quorum sensing in *Aeromonas hydrophila. Environ. Microbiol.* 4: 18–28.

Makin S.A. and Beveridge T.J. 1996. The influence of A-band and B-band lipopolysaccharide on the surface characteristics and adhesion of *Pseudomonas aeruginosa* to surfaces. *Microbiology* 142: 299–307.

Manefield M. and Turner S.L. 2002. Quorum sensing in context: Out of molecular biology and into microbial ecology. *Microbiology* 148: 3762–3764.

Mattmann M.E. and Blackwell H.E. 2010. Small molecules that modulate quorum sensing and control virulence in *Pseudomonas aeruginosa. J. Org. Chem.* 75: 6737–6746.

Merritt J., Qi F., Goodman S.D., Anderson M.H. and Shi W. 2003. Mutation of *luxS* affects biofilm formation in *Streptococcus mutans. Infect. Immun.* 71: 1972–1979.

Miller M.B. and Bassler B.L. 2001. Quorum sensing in bacteria. *Annu. Rev. Microbiol.* 55: 165–199.

Musthafa K.S., Sianglum W., Saising J., Lethongkam S. and Voravuthikunchai S.P. 2017. Evaluation of phytochemicals from medicinal plants of Myrtaceae family on virulence factor production by *Pseudomonas aeruginosa. APMIS* 125: 482–490.

Nadell C.D., Xavier J.B., Levin S.A. and Foster K.R. 2008. The evolution of quorum sensing in bacterial biofilms. *PLoS Biol.* 6: 171–179.

Nealson K.H. and Hastings J.W. 1979. Bacterial bioluminescence: Its control and ecological significance. *Microbiol. Rev.* 43: 496–518.

Neethirajan S., Clond M.A. and Vogt A. 2014. Medical biofilms- nanotechnology approaches. *J. Biomed. Nanotech.* 10: 1–22.

Ng W.L. and Bassler B.L. 2009. Bacterial quorum-sensing network architectures. *Annu. Rev. Genet.* 43: 197–222.

Njoroge J. and Sperandio V. 2009. Jamming bacterial communication: New approaches for the treatment of infectious diseases. *EMBO Mol. Med.* 1: 201–210.

Novick R.P. and Geisinger E. 2008. Quorum sensing in *Staphylococci. Annu. Rev. Genet.* 42: 541–564.

O'Toole G. and Stewart P. 2005. Biofilms strike back. *Nat. Biotechnol.* 23: 1378–1379.

O'Toole G.A., Kaplan H. and Kolter R. 2000. Biofilm formation as microbial development. *Annu. Rev. Microbiol.* 54: 49–79.

Otto M. 2004. Quorum-sensing control in *Staphylococci*—a target for antimicrobial drug therapy? *FEMS Microbiol. Lett.* 241: 135–141.

Palmer R.J. and Stoodley P. 2007. Biofilms 2007: Broadened horizons and new emphases. *J. Bacteriol.* 189: 7948–7960.

Pantanella F., Valenti P., Frioni A., Natalizi T., Coltella L. and Berlutti F. 2008. BioTimer Assay, a new method for counting *Staphylococcus* spp. in biofilm without sample manipulation applied to evaluate antibiotic susceptibility of biofilm. *J. Microbiol. Methods* 75: 478–484.

Pantanella F., Valenti P., Natalizi T., Passeri D. and Berlutti F. 2013. Analytical techniques to study microbial biofilm on abiotic surfaces: Pros and cons of the main techniques currently in use. *Ann Ig* 25: 31–42.

Papenfort K. and Bassler B.L. 2016. Quorum sensing signal-response systems in gram-negative bacteria. *Nat. Rev. Microbiol.* 14: 576–588.

Parsek M.R. and Greenberg E.P. 2005. Sociomicrobiology: The connections between quorum sensing and biofilms. *Trends Microbiol.* 13: 27–33.

Parsek M.R. and Greenberg E.P. 1999. Quorum sensing signals in development of *Pseudomonas aeruginosa* biofilms. *Methods Enzymol.* 310: 43–55.

Percival S.L., Malic S., Cruz H. and Williams D.W. 2011. Introduction to biofilms. In: Percival S.L. (Ed.) *Biofilms and Veterinary Medicine.* Springer Series on Biofilms 6. Springer-Verlag, Berlin, Germany.

Persson T., Givskov M. and Nielsen J. 2005. Quorum sensing inhibition: Targeting chemical communication in gram-negative bacteria. *Curr. Med. Chem.* 12: 3103–3115.

Pierson L.S. III, Keppenne V.D. and Wood D.W. 1994. Phenazine antibiotic biosynthesis in *Pseudomonas aureofaciens* 30-84 is regulated by phzR in response to cell density. *J. Bacteriol.* 176: 3966–3974.

Pikuta E.V. and Hoover R.B. 2007. Microbial extremophiles at the limits of life. *Crit. Rev. Microbiol.* 33: 183–209.

Piper K.R., Beck von Bodman S. and Farrand S.K. 1993. Conjugation factor of *Agrobacterium tumefaciens* regulates Ti plasmid transfer by autoinduction. *Nature* 362: 448–450.

Prakash B., Veeregowda B.M. and Krishnappa G. 2003. A survival strategy of bacteria. *Curr. Sci.* 85: 9–10.

Pratten J., Foster S.J., Chan P.F., Wilson M. and Nair S.P. 2001. *Staphylococcus aureus* accessory regulators: Expression within biofilms and effect on adhesion. *Microbes Infect.* 3: 633–637.

Prouty A.M., Schwesinger W.H. and Gunn J.S. 2002. Biofilm formation and interaction with the surfaces of gallstones by *Salmonella* spp. *Infect. Immun.* 70: 2640–2649.

Puskas A., Greenberg E.P., Kaplan S. and Schaefer A.L. 1997. A quorum-sensing system in the free-living photosynthetic bacterium *Rhodobacter sphaeroides. J. Bacteriol.* 179: 7530–7537.

Ramage G., Martinez J.P. and Lopez-Ribot J.L. 2006. Candida biofilms on implanted biomaterials: A clinically significant problem. *FEMS Yeast Res.* 6: 979–986.

Reading N.C. and Sperandio V. 2006. Quorum sensing: The many languages of bacteria. *FEMS Microbiol. Lett.* 254: 1–11.

Redfield R.J. 2002. Is quorum sensing a side effect of diffusion sensing? *Trends Microbiol.* 10: 365–370.

Roy R., Tiwari M., Donelli G. and Tiwari V. 2017. Strategies for combating bacterial biofilms: A focus on anti-biofilm agents and their mechanisms of action. *Virulence.* doi:10.1080/21505594.2017.1313372.

Rutherford S.T. and Bassler B.L. 2012. Bacterial quorum sensing: Its role in virulence and possibilities for its control. *Cold Spring Harb. Perspect. Med.* 2: a012427.

Sachs J.L., Mueller U.G., Wilcox T.P. and Bull J.J. 2004. The evolution of cooperation. *Q. Rev. Biol.* 79: 135–160.

Schauder S. and Bassler B.L. 2001. The languages of bacteria. *Genes Dev.* 15: 1468–1480.

Shapiro J.A. and Dworkin M. 1997. *Bacteria as Multicellular Organsims.* Oxford University Press, New York.

Sifri C.D. 2008. Quorum sensing: Bacteria talk sense. *Clin. Infect. Dis.* 47: 1070–1076.

Slater H., Alvarez-Morales A., Barber C.E., Daniels M.J. and Dow J.M. 2000. A two-component system involving an HD-GYP domain protein links cell–cell signalling to pathogenicity gene expression in *Xanthomonas campestris. Mol. Microbiol.* 38: 986–1003.

Smith W.A. 2005. Biofilms and antibiotic therapy: Is there a role for combating bacterial resistance by the use of novel drug delivery system? *Adv. Drug Deliv. Rev.* 57: 1539–1550.

Smith R.S. and Iglewski B.H. 2003. *Pseudomonas aeruginosa* quorum-sensing systems and virulence. *Curr. Opin. Microbiol.* 6: 56–60.

Sperandio V., Mellies J.L., Nguyen W., Shin S. and Kaper J.B. 1999. Quorum sensing controls expression of the type III secretion gene transcription and protein secretion in enterohemorrhagic and enteropathogenic *Escherichia coli. Proc. Natl. Acad. Sci. (USA)* 96: 15196–15201.

Srinivasan R., Devi K.R., Kannappan A., Pandian S.K. and Ravi A.V. 2016. *Piper betle* and its bioactive metabolite phytol mitigates quorum sensing mediated virulence factors and biofilm of nosocomial pathogen *Serratia marcescens* in vitro. *J. Ethnopharmacol.* 93: 592–603.

Srivastava S. and Bhargava A. 2014. Microbial biofilms: From nature to human body. In: Shukla D.S. and Pandey D.K. (Eds.) *Current Trend in Life Science.* JBC Press, New Delhi, India, pp. 1–16.

Srivastava S. and Bhargava A. 2016. Biofilms and human health. *Biotechnol. Lett.* 38: 1–22.

Srivastava S., Pathak N. and Srivastava P. 2011. Identification of limiting factors for the optimum growth of *Fusarium oxysporum* in liquid media. *Toxicol. Intern.* 18: 111–116.

Stoodley P., Sauer K., Davies D.G. and Costerton J.W. 2002. Biofilms as complex differentiated communities. *Annu. Rev. Microbiol.* 56: 187–209.

Suga H. and Smith K.M. 2003. Molecular mechanisms of bacterial quorum sensing as a new drug target. *Curr. Opin. Chem. Biol.* 7: 586–591.

Swift S., Downie J.A., Whitehead N.A., Barnard A.M.L., Salmond G.P.C. and Williams P. 2001. Quorum sensing as a population-density-dependent determinant of bacterial physiology. *Adv. Microb. Physiol.* 45: 199–270.

Taga M.E. and Bassler B.L. 2003. Chemical communication among bacteria. *Proc. Natl. Acad. Sci. (USA)* 100: 14549–14554.

Thomas M.S. and Wigneshwerareaj S. 2014. Regulation of virulence gene expression. *Virulence* 5: 832–834.

Uroz S., Oger P., Lepleux C., Collignon C., Frey-Klett P. and Turpault M.P. 2011. Bacterial weathering and its contribution to nutrient cycling in temperate forest ecosystems. *Res. Microbiol.* 162: 820–831.

van Bodman S.B., Willey J.M. and Diggle S.P. 2008. Cell-cell communication in bacteria: United we stand. *J. Bacteriol.* 190: 4377–4391.

Waters C.M. and Bassler B.L. 2005. Quorum sensing: Cell-to-cell communication in bacteria. *Ann. Rev. Cell Dev. Biol.* 21: 319–346.

Webb J.S., Givskov M. and Kjelleberg S. 2003. Bacterial biofilms: Prokaryotic adventures in multicellularity. *Curr. Opin. Microbiol.* 6: 578–585.

Wen Z.T. and Burne R.A. 2004. LuxS-mediated signaling in *Streptococcus mutans* is involved in regulation of acid and oxidative stress tolerance and biofilm formation. *J. Bacteriol.* 186: 2682–2691.

West S.A., Pen I. and Griffin A.S. 2002. Conflict and cooperation- cooperation and competition between relatives. *Science* 296: 72–75.

West S.A., Griffin A.S., Gardner A. and Diggle S.P. 2006. Social evolution theory for microorganisms. *Nat. Rev. Microbiol.* 4: 597–607.

Williams P., Camara M., Hardman A., Swift S., Milton D., Hope V.J., Winzer K., Middleton B., Pritchard D.I. and Bycroft B.W. 2000. Quorum sensing and the population-dependent control of virulence. *Philos. Trans. R. Soc. Lond. B: Biol. Sci.* 355: 667–680.

Winzer K. and Williams P. 2001. Quorum sensing and the regulation of virulence gene expression in pathogenic bacteria. *Intern. J. Med. Microbiol.* 291: 131–143.

Withers H.L. and Nordstrom K. 1998. Quorum-sensing acts at initiation of chromosomal replication in *Escherichia coli. Proc. Natl. Acad. Sci. (USA)* 95: 15694–15699.

Wu H., Moser C., Wang H.-Z., Høiby N. and Song Z.-J. 2015. Strategies for combating bacterial biofilm infections. *Intern. J. Oral Sci.* 7: 1–7.

Yarwood J.M. and Schlievert P.M. 2003. Quorum sensing in *Staphylococcus* infections. *J. Clin. Invest.* 112: 1620–1625.

Yarwood J.M., Bartels D.J., Volper E.M. and Greenberg E.P. 2004. Quorum Sensing in *Staphylococcus aureus* Biofilms. *J. Bacteriol.* 186: 1838–1850.

Zhang L., Murphy P.J., Kerr A. and Tate M.E. 1993. *Agrobacterium* conjugation and gene regulation by *N*-acyl-l-homoserine lactones. *Nature* 362: 446–448.

3 Biofuels
A Green Technology for the Future

Atul Bhargava and Francisco Fuentes

CONTENTS

3.1 INTRODUCTION

Energy is the most important input that drives economic development. An increasing demand for energy, highly volatile crude supply and prices, climate change, and widespread energy scarcity has been chiefly responsible for a search for alternative sources of energy that would be cheap and environmentally friendly. Biofuels fulfill these criteria and are fast becoming the fuels of the future. Biofuels or agrofuels are basically carbon-derived fuels (solid, liquid, or gaseous state), where the source of carbon is either plants or animals, and are therefore indirectly solar energy sources. Biofuels cover solid and liquid fuels, as well as various biogases, and include low-nutrient-input/high-per-acre-yield crops, agricultural or forestry waste, and other sustainable biomass feedstocks, including algae. Thus, biofuels comprise purpose-grown energy crops, as well as multipurpose plantations and by-products (residues and wastes). It is a renewable energy source, unlike other natural resources such as petroleum, coal, and nuclear fuels. Biofuels are termed green fuels because they are biodegradable, contribute to sustainability, emit less carbon dioxide, contain no sulfur, and decrease the global warming effect.

Biomass is essentially stored solar energy. Biofuels are fuels created from biomass through thermal conversion, chemical conversion, and biochemical conversion. Compared with most other fuels, the main feature of biofuels is renewability; that is, they can be produced again and again. This characteristic of biofuels makes them different from other energy sources commonly used today, such as nuclear fuels, coal, and petroleum. In 2008, global biofuel production reached about 83 billion liters, a more than fourfold increase as compared with production volumes in 2000. In 2010, biofuel production reached 105 billion liters, with the International Energy Agency (IEA) having an opinion that biofuels are likely to meet more than a quarter of world's demand for transportation fuels by the year 2050. The European Union is the world's largest biodiesel producer, whereas the United States and Brazil are the world's top ethanol producers, accounting together for 90% of global production.

3.2 HISTORY

Wood was the first form of biofuel that was used even by the ancient people for cooking and heating since prehistoric times, ever since man discovered fire. Apart from wood, solid biofuels such as dung and charcoal have been used since long and are still used today for heating, cooking, and other purposes in many developing countries. Whale oil was extensively used as the fuel of choice for lighting houses in the mid- and late-1700s. However, whale populations declined due to their killing, and consequently, the price of whale oil went up. Thereafter, a transition toward cheaper, fossil fuel-based kerosene occurred in the mid-1800s, after the modern method for refining kerosene was developed in 1846 by Abraham Gesner. The late seventeenth century saw the development of the early automotive engines utilizing steam power that gradually transformed to gasoline or petrol-fueled engines. Rudolph Diesel used peanut oil to power diesel compression engines in the late nineteenth century. Henry Ford, an early proponent of biofuels, built a factory that began making biofuels, but oil soon became the mainstream fuel of choice. Ford designed the Model T car, which was produced from 1903 to 1926, and it used hemp-derived ethanol as fuel. The Ford Motor Company used soybeans in various parts of the car, such as in the gearshift knobs and horn buttons. Hemp was used for making the body of the car as well as for fuel production. Henry Ford was the prominent supporter of the chemurgy movement during the 1920s and 1930s, which had ethanol as an important point on its agenda.

World War II led to a spurt in the demand of biofuels due to their increased use as an alternative for imported fuel. As compared with 60 million gallons' production during the World War I years, production of ethanol further increased during the World War II years to 600 million gallons. This period saw a rapid increase in the research on biofuel development, since several countries witnessed shortage of fuels. Germany underwent a serious shortage of fuels and developed the use of gasoline along with alcohol derived from potatoes. Similarly, Britain plagued by fuel shortages developed the concept of grain alcohol mixed with petrol. After World War II, cheap oil from the countries of Middle East Asia eased off the pressure. However, unrest and geopolitical tensions in the Middle East led to shortages and fueled fuel prices. The last 50 years saw the fuel prices swinging due to the following geopolitical tensions:

1. *1973 oil crisis*: Caused by export embargo by the Organization of Arab Petroleum Exporting Countries (OAPEC)

2. *1979 oil crisis*: Caused by the Iranian Revolution
3. *1990 oil crisis*: Caused by the Gulf War

The above-mentioned events, along with the rising prices of oil, emission of the greenhouse gases (GHG), and penchant for rural development, led to the renewed interest of academicians, researchers, and entrepreneurs in the development of biofuels.

3.3   TYPES OF BIOFUELS

Biofuels have been classified into different types per the classification given in the following. The following are considered the main types of biofuels:

1. *Biologically produced alcohols*: These are the types that utilize alcohols such as ethanol, methanol, propanol, and butanol produced through biomass. These are considered good sources of biofuels.
2. *Biologically produced gases*: Another type of biofuel is biogas, a gas produced by the anaerobic digestion of products, such as biodegradable wastes, or produced with the help of energy crops.
3. *Biologically produced oils*: These are oils such as straight vegetable oil (SVO), waste vegetable oil (WVO), and oils obtained from biological mass. These types of oils are usually used to run diesel engines.

Biofuels have also been divided in several ways by different agencies. All biofuels have been divided into conventional and advanced biofuels by the IEA.

1. *Conventional biofuels*: These are often referred to as first-generation biofuels and are mainly derived from land-based crops. Bioethanol produced by the fermentation and distillation of plants containing sugar or starch (e.g., sugarcane, sugar beet, wheat, and corn) and oil crop-based biodiesel produced from oilseeds (e.g., sunflower, soybean, rapeseed, castor, *Jatropha*, and palm) are included in this category.
2. *Advanced biofuels*: These are often referred to as second- or third-generation technologies. These technologies are still in research and development and include the use of inedible waste products and algae to produce biofuels.

However, the most followed classification of biofuels is based on the origin and production technology, according to which the biofuels have been divided into different generations. Biofuels have been broadly divided into three types:

1. *First-generation biofuels*: These are commercially produced using conventional technology such as fermentation, distillation, and transesterification with basic feedstocks such as seeds, animal fats, grains, and whole plants from crops such as sugar cane, maize, rapeseed, wheat, sunflower, and oil palm. These plants have primarily served as food or fodder and are still mainly used for edible purpose. The most common first-generation biofuels

are bioethanol, followed by biodiesel, vegetable oil, biogas, bioalcohols, green diesel, biofuel gasoline, bioethers, syngas, and solid biofuels such as wood, sawdust, domestic refuse, charcoal, agricultural waste, and dried manure. Ethanol is already a well-established industry, whereas commercialization efforts for butanol are ongoing.

2. *Second-generation biofuels*: These, also termed advanced biofuels, are produced from various types of biomass, viz. any source of organic carbon that is renewed rapidly as part of the carbon cycle. The second-generation biofuels are primarily made from cellulosic energy crops, agricultural forestry residues, or coproducts. The second-generation biofuels enable us to get rid of the fuel versus food dilemma by using all forms of lignocellulosic biomass instead of using only easily extractible sugars, starches, or oils, as in the first-generation biofuels. Lignocellulosic biomass is a composite structure of lignin, cellulose, and hemicellulose polymers. The second-generation biofuels are not yet produced on a large scale commercially, but a considerable number of pilot plants have been set up in recent years in North America, Europe, and Asia. Arkenol's pilot cellulosic ethanol plant established in Izumi, Japan, in 2002 uses mixed waste wood chips of cedar, pine, and hemlock and generates about 100–300 liters of ethanol per day. Iogen Corporation established a pilot demonstration facility in Ottawa, Canada, to produce fuel ethanol from lignocellulosic materials by using wheat straw as the feedstock. The first shipments of the cellulosic ethanol from the facility were delivered to the market in 2004. The Iogen facility has been designed and built to process 40 tons of wheat straw per day into ethanol, using enzymes made in its nearby enzyme-manufacturing facility. The SEKAB's pilot lignocellulosic ethanol plant in Örnsköldsvik, Sweden (funded by the European Union), was established in 2004. It produces 300–400 liters of ethanol per day.

3. *Third-generation biofuels*: These are based on algal biomass production. The algae are grown and harvested to extract oil, which can be converted into biodiesel through a process similar to the process for the production of the first-generation biofuels, or it can be refined into other fuels as replacements to petroleum-based fuels (Figure 3.1).

The third-generation biofuels are more energy-dense as compared with the first- and second-generation biofuels per area of harvest. The third-generation types are cultured as low-cost, high-energy, and renewable sources of energy. Production of biofuels from algae usually depends on their lipid content. The use of algae in biofuel production is advantageous, since it can grow in areas unsuitable for the first- and second-generation crops. This utilization of land would relieve stress on water and arable land used. The cultivation of microalgae in brackish water or on nonarable land may not incur land-use change, leading to minimal associated environmental impacts. Algae can be grown using sewage, wastewater, and saline water, which would help in preserving water for human consumption. Microalgae have such a high growth rate that they can double their biomass in periods as short as 3.5 h. In addition, many species have oil content in the

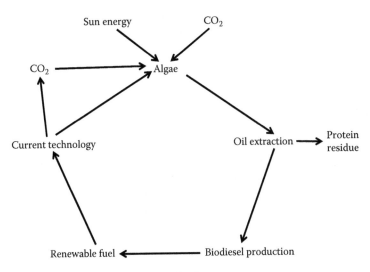

FIGURE 3.1 Third-generation biofuel production.

range of 20%–50% dry weight of biomass, with many producing valuable coproducts such as proteins and residual biomass after oil extraction that may be utilized as feed or fertilizer.

However, presently, the technology is uneconomical and nonsustainable, owing to low photon-to-fuel conversion efficiency (PFCE) of biodiesel production. This is attributed to enhanced biodiesel production, generally occurring under stress conditions, which reduces algal growth and biomass production. Efforts are on to overcome this drawback by recent advances in metabolic engineering of algae to increase lipid production, without compromising growth. This can be considered an important milestone toward sustainable biofuel production. Apart from this, a number of engineering approaches to increase algal fuel production, combination of algal biofuel production with production of high-value chemicals, the use of wastewater and/or sea water as culture media, and the development of more cost-effective bioreactors are currently being explored and are likely to make the algal biofuel production more profitable in the coming years. A recent approach describes the use of light-emitting diodes (LEDs) of different band wavelengths on the growth of microalgae in a closed, controlled environment to generate biomass and lipid yields. The exposure to LEDs of red wavelength post biomass generation leads to enhanced lipid production, with double lipid content harvested in 20 days' culture period.

4. *Fourth-generation biofuels*: The fourth-generation biofuels are based on photobiological solar fuels and electrofuels and envisage direct conversion of solar energy into fuel by using raw materials that are cheap, readily available, and renewable. However, these types are in the emerging stage, and

much needs to be done with respect to discovery of new-to-nature solutions and construction of synthetic living factories and designer microorganisms for efficient and direct conversion of solar energy to fuel.

3.4 BIOETHANOL

Ethanol is a colorless, biodegradable, and low-toxicity liquid. Bioethanol is mainly produced by the fermentation process, using different six carbon sugars as feedstocks and using microorganisms such as fungi, bacteria, and yeast. Starchy feedstocks such as sweet potato, potato, cassava, and cereal grains can also be utilized for bioethanol production by hydrolysis, followed by fermentation. In the sugar fermentation process, hydrolysis process breaks down the cellulosic part of the biomass into sugar solution, which can then be fermented into ethanol, as given in the following chemical reactions:

$$C_{12}H_{22}O_{11} + H_2O \xrightarrow{\text{Invertase}} C_6H_{12}O_6 + C_6H_{12}O_6$$

Sucrose   Fructose   Glucose

$$C_6H_{12}O_6 \xrightarrow{\text{Zymase}} C_2H_5OH + 2CO_2$$

Fructose/Glucose   Ethanol

Bioethanol has a number of benefits over conventional fuels. It is biodegradable, less toxic than fossil fuels, and can help reduce the amount of carbon monoxide produced by older engines, thus improving air quality. Another advantage of bioethanol is the ease with which it can be integrated into the existing road transport fuel system, with less of engine modifications required. The most common blend is 10% ethanol and 90% petrol (E10). Vehicle engines require no modifications to run on E10, and vehicle warranties are also unaffected.

The commercial feasibility of bioethanol mainly depends on its production costs and the policy framework adopted by governments. Brazil produces about 59% of world ethanol and is the lowest cost producer of ethanol, which is due to lower input costs, large and efficient plants, and use of sugarcane as feedstock.

3.5 BIODIESEL

For automobile usage, diesel is a mixture of C9–C23 hydrocarbons, with an average carbon length of 16. Linear, branched, and cyclic alkanes compose 75% of the fuel mixture, whereas the rest 25% is composed of aromatics. Biodiesel is derived from vegetable oils (e.g., rapeseed oil, *Jatropha*, soy, and palm oil) by the reaction of triglycerides of the fatty acid with methanol under the basic conditions (e.g., sodium hydroxide as the catalyst) to yield the methyl ester of the fatty acid (biodiesel) (transesterification reaction). Transesterification is the process of reacting a triglyceride (oil) with an alcohol in the presence of a catalyst, such as sodium hydroxide and potassium hydroxide, to chemically break the molecule of the oil into methyl or ethyl esters (Figure 3.2). Transesterification

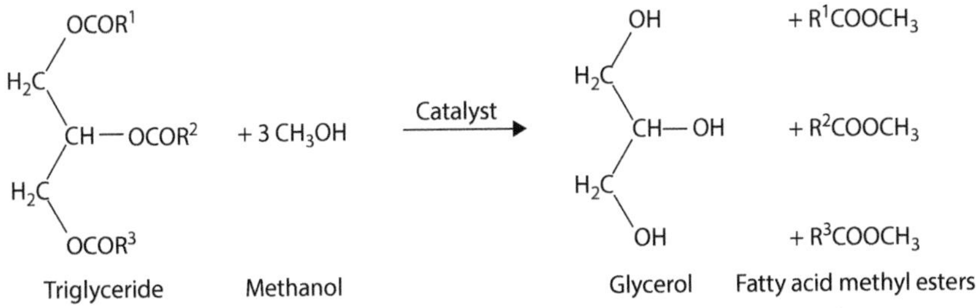

FIGURE 3.2 A transesterification reaction.

consists of various reversible reactions, wherein a triglyceride is converted step-wise into diglyceride, monoglyceride, and finally glycerol.

Biodiesel can also be obtained from vegetable oils via their hydrocracking, which yields a diesel mainly composed of alkanes, and is similar to petroleum diesel. Although large amounts of low-cost oils and fats that can be converted to biodiesel are available, most of the biodiesel is currently made from edible oils by using methanol and an alkaline catalyst. Low-cost oils and fats are less economical, since a two-step esterification process is required, as they often contain large amounts of free fatty acids that cannot be converted to biodiesel by using an alkaline catalyst. Biodiesel can either be burnt directly in diesel engines or blended with diesel derived from fossil fuels. Biodiesel is much more superior in comparison with petroleum-based diesel because of its effect on emissions, its cetane number, its flash point, and its lubricity. Another advantage of biodiesel is that it is nontoxic and can be used directly in diesel engines without modifications. Since biodiesel is physically similar to petroleum-based diesel fuel, it can be blended with diesel fuel in any proportion. The following are the most common commercial forms of biodiesel:

- *B100*: 100% biodiesel
- *B20*: Mixture consisting of 20% biodiesel and 80% petroleum diesel
- *B2*: Mixture consisting of 2% biodiesel and 98% petroleum diesel

3.6 ADVANCED BIOFUELS

Advanced biofuels are biofuels that have high energy content and are compatible with storage and transportation infrastructures designed for petroleum-based products. These are also economically feasible to produce on a commercial scale. Butanol, isobutanol, propanol, and isopentanol are some of the recently developed advanced biofuels. Advanced biofuels, such as long-chain alcohols and isoprenoid- and fatty acid-based biofuels, have physical properties that more closely resemble petroleum-derived fuels, and as such, they are an attractive alternative for the future supplementation or replacement of petroleum-derived fuels.

The engineering of microorganisms appears to be the most convenient and cost-effective approach for the large-scale production of advanced biofuels. This is on account of four developments:

1. Rapid engineering of microbial biosynthetic pathways is possible by utilizing advances in molecular systems and synthetic biology. This allows for the production of a variety of advanced biofuel candidates from the fatty acid and isoprenoid pathways.
2. The microbial production of advanced biofuels is aided by the application of industrial fermentation knowledge.
3. It has become possible to produce advanced microbial biofuels in bioreactors, and the production facilities can be placed wherever needed.
4. Microbes could generate biofuel from starchy agricultural products as well as from lignocellulosic biomass that cannot be used for food once the technology is developed for the breakdown of lignocellulosic biomass.

Peralta-Yahya and Keasling (2010) have discussed about the exploitation of several metabolic pathways for the production of known and potential advanced biofuels. These are as follows:

1. Heterologous expression of the Clostridial C3–C4 biosynthetic pathway for the production of isopropanol and 1-butanol in *Escherichia coli* and *Saccharomyces cerevisiae*.
2. Generation of higher alcohols through rerouting of the amino acid biosynthetic pathway.
3. Production of isoprenoid-based biofuels through metabolic engineering of the isoprenoid biosynthetic pathway.
4. Production of fatty acid-based biofuels by metabolic engineering of the fatty acid biosynthetic pathway.

The last decade has led to the development of bioelectrochemical cells (BEC); notably among them are the microbial fuel cells (MFCs) and microbial electrolysis cells (MECs), which utilize organic biomass and wastewaters and have been extensively exploited for bioelectricity and biohydrogen production. Both the above-mentioned cells function on the similar principle and use common microorganisms for bioenergy production. However, the BECs are in their infancy, and a better understanding of the microbial pathways such as electroactive biofilm formation, electron transfer mechanisms, and their modifications can help improve the energy output from these systems, which is critical in improving the bioenergy production.

Presently, most of the improvements in the production of advanced biofuels have been achieved by fine-tuning discrete steps within the production pathway or by regulating the carbon flux to and from the pathway. Functional genomic techniques enable us to simultaneously monitor thousands of parameters and accurately represent a snapshot of the cell metabolism by integrating these system-wide data into models. The initial step envisages engineering the microorganism for biofuel production, and then, the engineered microbe is profiled using functional genomics and metabolic flux analysis to identify the potential bottlenecks in the production and the toxicities resulting from pathway expression. The final step involves the incorporation of systems biology predictions into the microbes and testing of fuel production. Synthetic biology advances these aims further by offering a less resource-intensive and more time-efficient alternative with which to achieve improved biofuel production. During microbial production of advanced biofuels, accumulation of the final product can negatively impact the cultivation of the host microbe and limit the production levels. Therefore, improving solvent tolerance has become an integral part of engineering microbial production strains for biofuel production. Mechanisms ranging from chaperones to transcriptional factors have been used to obtain solvent-tolerant strains.

3.7   ADVANTAGES OF BIOFUELS

One of the main objectives of developing biofuels is sustainability. The sustainability concept is based on the three pillars of economic, social, and environmental sustainability. In economic terms, biofuels have to be cost-effective and competitive against conventional fuels. In social terms, biofuels development can remodel the agricultural economy and create a massive new demand in the agricultural sector. Sustainability criteria also demonstrate how different forms of energy perform in terms of energy efficiency and environment impact.

1. *Social benefits*: Firewood, the most common fuel in rural areas in many developing countries, has a negative effect on health and adverse impact on people's quality of life, as people spend a considerable amount of time collecting it. Women, older people, and children suffer disproportionately because of their relative dependence on traditional fuels and their exposure to smoke from cooking, which happens to be the main cause of respiratory diseases in such societies. Access to modern decentralized small-scale energy technologies, particularly renewables (including biofuels), are an important element for successful and effective poverty-alleviation policies. Large deforested areas could be recovered by planting of crops that could produce biofuels, which in turn would also have a significant impact on employment opportunities, mainly in the rural areas. Since the production of biofuels requires crops and other plant-/animal-based matter, agrarian economies, such as those of the developing countries, will have a distinct advantage over others in biofuel production. Biofuels are likely to develop a new rural industry, increase economic activity, create more jobs, and improve income generation through labor-intensive agriculture.

2. *Energy security*: Constant high demand in oil has raised the oil prices and also caused supply problems. Biofuels reduce the dependence of countries over other countries for fuels, since each one is capable of producing biofuels. Fuel independence would definitely increase domestic security and self-dependence among such economies. Biofuels can reduce the burden

of energy imports for energy-deficient countries and could contribute to addressing energy imbalances around the world (Von Braun and Pachauri, 2006). However, many believe that the potential of biofuels replacing fossil fuels is much hyped. The IEA foresees that the percentage of biofuel used for transportation would be 2.3% by 2015 and 3.2% by 2030. This is quite small and much lower than the expectations of the public, fed by the glowing stories about biofuels all around.

3. *Cleaner environment*: Biofuels are biodegradable and not harmful when released in the environment. Biofuels are considered much more environmentally friendly than fossil fuels, because biofuels significantly reduce GHG emissions compared with fossil fuels and as a result reduce pollution. GHG emissions and energy requirements of biofuel production vary widely, depending on the feedstock, cultivation method, conversion technologies, and energy-efficiency assumptions. The greatest reduction in GHG emissions can be derived from sugarcane-based bioethanol and the second-generation biofuels such as lignocellulosic bioethanol and Fischer–Tropsch biodiesel. In contrast, maize-derived bioethanol shows the worst GHG emission performance, where, in some cases, the GHG emissions can be even higher than those related to fossil fuels. Mixing ethanol with fossil fuel to make gasoline helps reduce air pollution, because it causes less sulfur oxide, lead, and other polluting particles to be released in the air when the fuel is burned. Studies that have compared the cellulosic liquid biofuel production from conventional petroleum use and first-generation food crop-based biofuel production have shown significant reductions in GHG emissions. Zah et al. (2007) observed about 80% lower GHG emissions from biofuels as compared with fossil fuels. Some biofuels have GHG savings greater than 100% due to the cogeneration of products. For example, in the case of biomethane from manure, GHG benefits of up to 174% have been reported on using feedstocks from waste, whereas especially methane escaping from biogas plants may lower the overall performance to only 37% GHG savings. Thus, the GHG balance of biofuels for transport depends on several factors. The cogeneration of other products may lead to improved performance, whereas intensive agricultural production and conversion of natural land to cropland may lead to negative results.

4. *Renewable feedstock resource*: Apart from producing biofuels, biomass will also provide a renewable feedstock for the chemicals industry, replacing petrochemicals. The use of biomass as a feedstock has had significant impact on the chemical industry with respect to solvents, plastics, lubricants, and fragrances.

3.8   DISADVANTAGES/RISKS OF BIOFUELS

Biofuels have received a lot of attention in recent decades for reducing fossil-fuel dependency and GHG emissions. However, biofuels can trigger land-use change, resulting in unwanted social and environmental effects, such as loss of biodiversity, natural vegetation, labor competition, biofuels-induced food price increases, and overexploitation of water resources. Some of the negative aspects of use of biofuels are discussed as follows:

1. *Food shortages and food prices*: There has been a lot of concern about the impacts of biofuel on food availability and prices. The greater demand for biofuels creates incentives to reallocate resources, notably land and water from food to fuel production. The relationship between biofuel and food is a very complicated one. If biofuels become lucrative for farmers, they tend to grow crops strictly for biofuel production, which will result in less food in the market. Less food means higher prices, rise in inflation, and more hungry people worldwide. Rising food prices hurt consumers in general but hurt poorer consumers in particular, because the poor tend to spend greater percentages of their income on food items. This is the reason why many people are against the production of the first-generation biofuels, because food production must be far more important than fuel production, regardless of high profits. Guatemala, a green country with the fourth highest malnutrition rate in the world, is a classic example of the impact of biofuel production on livelihood. Many families practicing subsistence agriculture were evicted to expand sugarcane and African palm plantation, catering to biofuel production. Food crops are being replaced to make way for fuel crops, leading to skyrocketing of the food prices. This has led to hunger and malnutrition in many of the evicted families, who were earlier leading a comfortable life.

2. *Loss in biodiversity*: The demand for land needed for food, animal feed, and biofuel production has rapidly increased, leading to an increased pressure on land and other resources, such as water. Biofuels production will create biodiversity problems because many animals will lose their habitats, as more and more land will be used for biofuels production. It could also cause even bigger deforestation problems in some developing countries, as without sustainable management, forests could be cleared to make way for biofuels production. Apart from this, with any agricultural practice, there are risks associated with the exclusive culture (monoculture) of biofuel sources. However, this can be avoided by good management practices. The use of the second-generation biofuel sources such as native grasses in mixed plantations can improve biodiversity.

3. *Energy imbalance*: If the total of the world's crops is directed to produce biofuels, this would constitute between 9% and 13% of the world's primary energy. By 2050, this would correspond to only 4%–6% of the world's energy, while mobilizing 85% of the world's freshwater resources. The above-mentioned estimates highlight a gross imbalance between resource utilization and energy generation. When we consider the social and environmental costs such as insufficient food to feed a growing global population, higher emissions, and a greater degradation of our ecosystems, the opinion about biofuels needs to be revised.

4. *Water quality and shortages*: If there is an increase in the cultivation and use of crops such as maize for bioethanol production, it may harm water

quality due to excessive use of inputs such as fertilizers and pesticides. The rapidly growing bioenergy crops consume more water than natural vegetation or other food crops and consequently may lead to water shortages in the future. For example, sugarcane, a raw material for ethanol production, requires large amounts of water, which may become problematic in semi-arid areas. An evaluation of the water use for biofuel production conducted by the Environmental Defense for the Ogallala Aquifer (which is under eight western U.S. states, ranging from Nebraska to Texas) has led to the fear that misuse of the aquifer could possibly lead to a reoccurrence of the *Dust Bowl* conditions of the 1930s. The ethanol plants are expected to require as much as 2.6 billion gallons of water to process corn and convert it to fuel. The processing of some feedstocks requires large volumes of water and tends to generate effluent. Thus, the environment into which we seek to place the new technology must be scrutinized to avoid the loss of water supplies and other resources. The introduction and enforcement of appropriate technologies, regulations, and standards can help reduce some of these problems.

3.9   TRANSGENIC TECHNOLOGY IN BIOFUEL PRODUCTION

During the domestication and selection process, important traits helpful in biofuel production, such as low lignin content and high cellulose, were lost systematically from the gene pool. The advent of transgenic technology has enabled us to modify plant architecture vis-à-vis biofuel production by making available the biofuel productivity traits or pathways in a cloned form, which leads to integration of transgenes in a highly directed manner. For efficient biofuel production, technologies can be designed to effectively break down complex carbohydrates into simple sugars. This can be achieved by biotechnologically engineering transgenic crops having modified lignin and cellulosic biosynthetic pathway, which will make their decomposition into biofuels more efficient and rapid when acted upon by microbes. Plants can be genetically engineered to more quickly degrade plant parts such as wheat and paddy straw, stalks of cotton, and corn, which are normally discarded during threshing. This will not only enhance the rate of production of biofuels but also improve the efficiency of the whole production process. The transgenic approach toward biofuels can be in the following three directions:

1. Manipulating genes and pathways involved in starch and sugar metabolism
2. Manipulating genes and pathways involved in cellulose and lignin biosynthesis
3. Manipulating genes and pathways involved in lipid metabolism

Another approach is to isolate microbes that can survive in hydrocarbon-rich environments and transfer the tolerance mechanisms to a suitable production strain. A well-studied organism with good genetic tools available could serve as an ideal host that can be engineered for both biofuel production and tolerance.

3.10   THE FUTURE OF BIOFUELS

The interest in biofuels has increased significantly amid fluctuating crude oil prices, with ample evidence to suggest that the environmental impacts and the inefficiencies in biofuel production can be avoided through the use of appropriate biomass and production method. Automobile manufacturers and the aviation sector have started exploiting biofuels and could contribute a lot in the popularity and greater usage of this new-generation fuel. The U.S. Air Force has already approved a 50% biofuel blend in its F-15 and F-16 fighter jets and C-17 transport planes. The U.S. Navy hopes to be at 50% use of non-fossil-fuel energy by 2020. In the near future, advanced biofuels need to have properties very similar to current transportation fuels, which would allow for maximum compatibility with existing engine design, storage infrastructure, and distribution systems.

A number of technological advances are needed to improve the benefits of biofuels (Ragauskas et al., 2006). Some of these are enumerated as follows:

1. Increased yield of plants, especially biomass
2. Lower agricultural inputs such as fertilizers, agrochemicals, and irrigation
3. Improved management of soil health under intensive agriculture
4. Increased ethanol yield during fermentation processes
5. Development of large-scale processes for the production of biodiesel from lower cryptogams such as algae
6. Development of efficient compression combustible engines to avoid the esterification step and use plant oils directly
7. Less energy input from fossil fuel by improvement in processing
8. Improved vehicle efficiency for better utilization

The outlook for biofuel development and usage will depend on a number of interrelated factors, namely the price of crude oil, availability of low-cost feedstocks, supportive policies by governments, technological breakthroughs, and competition from unconventional fossil fuel alternatives. Technological advances, higher biomass yields per acre, and more gallons of biofuel per ton of biomass could steadily reduce the economic cost and environmental impacts of biofuel production. Biofuel production will likely be more profitable and environmentally friendly in tropical regions, where growing seasons are longer, per acre biofuel yields are higher, and fuel and other input costs are lower.

The future of the biofuels rests on the ability to exploit the potential of this fuel source without compromising agricultural land use or exacerbating climate change, that is, sustainable biofuel production. A sustainable biofuel production system is one that is economically viable, conserves the natural resource base, and ensures social well-being. Biofuels are expected to reduce GHG emissions; lower the dependence on petroleum, which is associated with political and economic vulnerability; and revitalize the economy by increasing the demand and prices of agricultural products. From a sustainability perspective, biofuels offer several benefits and risks. On the brighter side, biofuels can contribute to increased energy security, reduce GHG emissions, spur growth in rural areas, and significantly improve the air quality. Regarding the potential risks, expansion of biofuels under intensive agricultural systems can lead to negative impacts on biodiversity and water availability and its quality, soil degradation, negative carbon and energy balances, conflict with food production, undermining of food security, and worsening of the GHG emission levels due to indirect land-use change. There is a need to balance the economic benefits with environmental and social impacts.

BIBLIOGRAPHY

Alper H. and Stephanopoulos G. 2009. Engineering for biofuels: Exploiting innate microbial capacity or importing biosynthetic potential? *Nat. Rev. Microbiol.* 7: 715–723.

Atsumi S., Hanai T. and Liao J.C. 2008. Non-fermentative pathways for synthesis of branched-chain higher alcohols as biofuels. *Nature* 451: 86–89.

Brennana L. and Owendea P. 2010. Biofuels from microalgae—A review of technologies for production, processing, and extractions of biofuels and coproducts. *Renew. Sustain. Energy Rev.* 14: 557–577.

Bungay H.R. Biomass refining. *Science* 218: 643–646.

Charles M.B., Ryan R., Ryan N. and Oloruntoba R. 2007. Public policy and biofuels: The way forward? *Energy Policy* 35: 5737–5746.

Chisti Y. 2008. Biodiesel from microalgae beats bioethanol. *Trends Biotechnol.* 26: 126–131.

Coelho S.T., Lucon O. and Guardabassi P. 2005. Biofuels: Advantages and trade barriers. *United Nations Conference on Trade and Development*, New York.

Coyle W. 2007. The future of biofuels: A global perspective. *Amber Waves* 5: 24–29.

de Fraiture C., Giordano M. and Liao Y. 2008. Biofuels and implications for agricultural water use: Blue impacts of green energy. *Water Policy* 10 (S1): 67–81.

de Gorter H. and Just D.R. 2010. The social costs and benefits of biofuels: The intersection of environmental, energy and agricultural policy. *Appl. Econ. Pers. Pol.* 32: 4–32.

De Oliveria M.E.D., Vaughan B.E. and Rykiel E.J. 2005. Ethanol as fuel: Energy, carbon dioxide balances and ecological footprint. *BioScience* 55: 593–602.

Demirbas M.F. 2006. Current technologies for biomass conversion into chemicals and fuels. *Energy Sour. Part A* 28: 1181–1188.

Demirbas A. 2006. Global biofuel strategies. *Energy Edu. Sci. Technol.* 17: 27–63.

Demirbas A. 2008. *Biodiesel: A Realistic Fuel Alternative for Diesel Engines.* Springer, London, UK.

Demirbas A. 2009. Political, economic and environmental impacts of biofuels: A review. *Appl. Energy* 86: S108–S117.

Doornbosch R. and Steenblik R. 2007. Biofuels: Is the cure worse than the disease? Retrieved from http://www.oecd.org/dataoecd/15/46/39348696.pdf.

Dunlop M.J. 2011. Engineering microbes for tolerance to next generation biofuels. *Biotechnol. Biofuels* 4: 32.

EASAC (European Academies Science Advisory Council). 2012. The current status of biofuels in the European Union, their environmental impacts and future prospects. EASAC Policy Report 19.

EASAC (European Academies Science Advisory Council). 2013. Planting the future: Opportunities and challenges for using crop genetic improvement technologies for sustainable agriculture. EASAC Policy Report 21.

Eckermann E. 2001. *World History of the Automobile.* SAE Press, Warrendale, PA.

Elbehri A., Segerstedt A. and Liu P. 2013. *Biofuels and the Sustainability Challenge.* Food and Agricultural Organization, Rome, Italy.

Escobar J., Lora E., Venturini O., Yáñez E., Castillo E. and Almazan O. 2009. Biofuels: Environment, technology and food security. *Renew. Sustain. Energy Rev.* 13: 1275–1287.

FAO (Food and Agricultural Organization). 2000. The energy and agriculture nexus. Environment and Natural Resource Working Paper 4, Annex 1. Retrieved from http://www.fao.org/docrep/003/X8054E/x8054e00.HTM.

Fargione J., Hill J., Tilman D., Polasky S. and Hawthorne P. 2008. Land clearing and the biofuel carbon debt. *Science* 319: 1235–1238.

Farrel A.E., Plevin R.J., Turner B.T., Jones A.D., O'Hare M. and Kammen D.M. 2006. Ethanol can contribute to energy and environmental goals. *Science* 311: 506–508.

Furtado A., Lupo J.S., Hoang N.V., Healey A., Singh S., Simmons B.A. and Henry R.J. 2014. Modifying plants for biofuel and biomaterial production. *Plant Biotechnol. J.* 12: 1246–1258.

Gaffney J.S. and Marley N.A. 2009. The impacts of combustion emissions on air quality and climate—from coal to biofuels and beyond. *Atmos. Environ.* 43: 23–36.

Gamborg C., Millar K., Shortall O. and Sandøe P. 2012. Bioenergy and land use: Framing the ethical debate. *J. Agric. Environ. Ethics* 25: 909–925.

Glaser J.A. 2008. Biofuel point/counterpoint. *Clean Techn. Environ. Policy* 10: 113–116.

Goldemberg J. and Guardabassi P. 2009. Are biofuels a feasible option? *Energy Policy* 37: 10–14.

Gomez L.D., Clare G.S. and McQueen-Mason J. 2008. Sustainable liquid biofuels from biomass: The writing's on the walls. *New Phytol.* 178: 473–485.

Gouveia L. 2011. *Microalgae as a Feedstock for Biofuels.* Springer Briefs in Microbiology. Springer-Verlag, Berlin, Germany.

Groom M., Gray E. and Townsend P. 2007. Biofuels and biodiversity: Principles for creating better policies for biofuel production. *Conserv. Biol.* 22: 602–609.

Hall D.O. 1991. Biomass energy. *Energy Policy* 19: 711–737.

Hanna M.A., Isom L. and Campbell J. 2005. Biodiesel: Current perspectives and future. *J. Sci. Ind. Res.* 64: 854–857.

Hansen A.C., Zhang Q. and Lyne P.W.L. 2005. Ethanol-diesel fuel blends—A review. *Biores. Technol.* 96: 277–285.

Haughton A.J., Bond A.J., Lovett A.A., Dockerty T., Sünnenberg G., Clark S.J., Bohan D.A. et al. 2009. A novel, integrated approach to assessing social, economic and environmental implications of changing rural land-use: A case study of perennial biomass crops. *J. Appl. Ecol.* 46: 323–333.

Havlík P., Schneider U.A., Schmid E., Böttcher H., Fritz S., Skalský R., Aoki K. et al. 2010. Global land-use implications of first and second generation biofuel targets. *Energy Policy* 39: 5690–5702.

Ho S.-H., Ye X., Hasunuma T., Chang J.-S. and Kondo A. 2014. Perspectives on engineering strategies for improving biofuel production from microalgae: A critical review. *Biotechnol. Adv.* 32: 1448–1459.

House of Commons Environmental Audit Committee. 2008. Are biofuels sustainable? Retrieved from http://www.publications.parliament.uk/pa/cm200708/cmselect/cmenvaud/76/76.pdf.

IEA (International Energy Agency). 2010. Sustainable production of second-generation biofuels. Potential and perspectives in major economies and developing countries. Retrieved from http://www.iea.org/publications/freepublications/publication/biofuels_exec_summary.pdf.

Jojima T., Inui M. and Yukawa H. 2008. Production of isopropanol by metabolically engineered *Escherichia coli*. *Appl. Microbiol. Biotechnol.* 77: 1219–1224.

Kalscheuer R., Stolting T. and Steinbuchel A. 2006. Microdiesel: *Escherichia coli* engineered for fuel production. *Microbiology* 152: 2529–2536.

Karp A., Haughton A.J., Bohan D.A., Lovett A.A., Bond A.J., Dockerty T., Sunnenberg G. et al. 2009. Perennial energy crops: Implications and potential. In: *What is Land For? The Food, Fuel and Climate Change Debate.* Earthscan, London, UK, pp. 47–72.

Karp A. and Richter G.M. 2011. Meeting the challenge of food and energy security. *J. Exp. Bot.* 62: 3263–3271.

Katha S. 2006. Bioenergy and agriculture: Promises and challenges. Environmental Effects of Bioenergy. International Food Policy Research Institute, Washington, DC. Brief 4 of 12.

Kumar R. and Kumar P. 2017. Future microbial applications for bioenergy production: A perspective. *Front. Microbiol.* 8: 450.

Ladygina N., Dedyukhina E.G. and Vainshtein M.B. 2006. A review on microbial synthesis of hydrocarbons. *Process Biochem.* 41: 1001–1014.

Lee S.K., Chou H., Ham T.S., Lee T.S. and Keasling J.D. 2008. Metabolic engineering of microorganisms for biofuels production: From bugs to synthetic biology to fuels. *Curr. Opin. Biotechnol.* 19: 556–563.

Levidow L. and Paul H. 2008. *Land-use, Bioenergy and Agro-biotechnology.* WBGU, Berlin, Germany.

Liao J.C., Mi L., Pontrelli S. and Luo S. 2016. Fuelling the future: Microbial engineering for the production of sustainable biofuels. *Nat. Rev. Microbiol.* 14: 288–304.

Ling H., Teo W., Chen B., Leong S.S. and Chang M.W. 2014. Microbial tolerance engineering toward biochemical production: From lignocellulose to products. *Curr. Opin. Biotechnol.* 29: 99–106.

Logan B.E., Wallack M.J., Kim K.Y., He W., Feng Y. and Saikaly P.E. 2015. Assessment of microbial fuel cell configurations and power densities. *Environ. Sci. Technol. Lett.* 2: 206–214.

Maity J.P., Bundschuh J., Chen C.-Y. and Bhattacharya P. 2014. Microalgae for third generation biofuel production, mitigation of greenhouse gas emissions and wastewater treatment: Present and future perspectives: A mini review. *Energy* 78: 104–113.

Markevičius A., Katinas V., Perednis E. and Tamašauskienė M. 2010. Trends and sustainability criteria of the production and use of liquid biofuels. *Renew. Sustain. Energy Rev.* 14: 3226–3231.

Medipally S.R., Yusoff F.M., Banerjee S. and Shariff M. 2015. Microalgae as sustainable renewable energy feedstock for biofuel production. *BioMed Res. Int.* 2015: 13p. doi:10.1155/2015/519513.

Mohr A. and Raman S. 2013. Lessons from first generation biofuels and implications for the sustainability appraisal of second generation biofuels. *Energy Policy* 63: 114–122.

Mukhopadhyay A. 2015. Tolerance engineering in bacteria for the production of advanced biofuels and chemicals. *Trends Biotechnol.* 23: 498–508.

Mukhopadhyay A., Redding A., Rutherford B. and Keasling J.D. 2008. Importance of systems biology in engineering microbes for biofuel production. *Curr. Opin. Biotechnol.* 19: 228–234.

Murphy R., Woods J., Black M. and McManus M. 2011. Global developments in the competition for land from biofuels. *Food Policy* 36: S52–S61.

Nag A. 2007. *Biofuels Refining and Performance.* McGraw-Hill, New York.

Obersteiner M. 2010. Global land-use implications of first and second generation biofuel targets. *Energy Policy* 39: 5690–5702.

Oxfam. 2007. Bio-fuelling poverty: Why the EU renewable-fuel target may be disastrous for poor people. Oxfam Briefing Note. Retrieved from http://www.oxfam.org.nz/imgs/pdf/biofuels%20briefing%20note.pdf.

Panbdey A., Lee D.-J., Chisti Y. and Soccol C.R. 2014. *Biofuels from Algae.* Elsevier, Amsterdam, the Netherlands.

Peralta-Yahya P.P. and Keasling J.D. 2010. Advanced biofuel production in microbes. *Biotech. J.* 5: 147–162.

Peskett L., Slater R., Stevens C. and Dufey A. 2007. Biofuels, agriculture and poverty reduction. In: Farrington J. (Ed.), *Natural Resources Perspectives*. Overseas Development Institute, London, UK.

Pilgrim S. and Harvey M. 2010. Battles over biofuels in Europe: NGOs and the politics of markets. *Sociological Research Online*, 15. Retrieved from http://www.socresonline.org.uk/15/3/4.html.

Rabaey K. and Rozendal R.A. 2010. Microbial electrosynthesis: Revisiting the electrical route for microbial production. *Nat. Rev. Microbiol.* 8: 706–716.

Ragauskas A.J., Williams C.K., Davison B.H., Britovsek G., Cairney J., Eckert C.A., Frederick W.J. et al. 2006. The path forward for biofuels and biomaterials. *Science* 311: 484–489.

Rajagopal D. and Zilberman D. 2007. Review of environmental, economic and policy aspects of biofuels. Policy Research Working Paper WPS4341. The World Bank Development Research Group, Washington, DC.

Rajagopal D., Sexton S., Roland-Holst D. and Zilberman D. 2007. Challenge of biofuel: Filling the tank without emptying the stomach? *Environ. Res. Lett.* 2: 1–9.

Rude M. and Schirmer A. 2009. New microbial fuels: A biotech perspective. *Curr. Opin. Microbiol.* 12: 274–281.

Ruth L. 2008. Bio or bust? The economic and ecological cost of biofuels. *EMBO Rep.* 9(2): 130–133.

Schubert C. 2006. Can biofuels finally take center stage? *Nat. Biotechnol.* 24: 771–784.

Searchinger T. 2011. How biofuels contribute to the food crisis. *Washington Post*.

Sengers F., Raven R.P.J.M. and Van Venrooij A. 2010. From riches to rags: Biofuels, media discourses, and resistance to sustainable energy technologies. *Energy Policy* 38: 5013–5027.

Severes A., Hegde S., D'Souza L. and Hegde S. 2017. Use of light emitting diodes (LEDs) for enhanced lipid production in micro-algae based biofuels. *J. Photochem. Photobiol. B.* 170: 235–240.

Sims R.E.H., Mabee W., Saddler J.N. and Taylor M. 2010. An overview of second generation biofuel technologies. *Bioresour. Technol.* 101: 1570–1580.

Singh A., Pant D., Korres N.E., Nizami A.S., Prasad S. and Murphy J.D. 2010. Key issues in life cycle assessment of ethanol production from lignocellulosic biomass: Challenges and perspectives. *Bioresour. Technol.* 101: 5003–5012.

Sorda G., Banse M. and Kemfert C. 2010. An overview of biofuel policies across the world. *Energy Policy* 38: 6977–6988.

Thornley P. and Gilbert P. 2013. Biofuels: Balancing risks and rewards. *Interface Focus* 3: 2042–8901.

UNEP (United Nations Environment Programme). 2009. Towards sustainable production and use of resources: Assessing Biofuels. UNEP, Division of Technology Industry and Economics, Paris, France.

van der Horst D. and Vermeylen S. 2011. Spatial scale and social impacts of biofuel production. *Biomass Bioenergy* 35: 2435–2443.

Von Braun J. and Pachauri R.K. 2006. The promises and challenges of biofuels for the poor in developing countries. International Food Policy Research Institute 2005–2006 Annual Report Essay.

Wackett L.P. 2008. Biomass to fuels via microbial transformations. *Curr. Opin. Chem. Biol.* 12: 1–7.

Wassenaar T. and Kay S. 2008. Biofuels: One of many claims to resources. *Science* 321: 201.

Wijffels R.H., Kruse O. and Hellingwerf K.J. 2013. Potential of industrial biotechnology with cyanobacteria and eukaryotic microalgae. *Curr. Opin. Biotech.* 24: 405–413.

Williams P.J. 2007. Biofuel: Microalgae cut the social and ecological costs. *Nature* 450: 478.

Yan Y. and Liao J.C. 2009. Engineering metabolic systems for production of advanced fuels. *J. Ind. Microbiol. Biotechnol.* 36: 471–479.

Zah R., Böni H., Gauch M., Hischier R., Lehmann M. and Empa P.W. 2007. Ökobilanz von Energieprodukten: Ökologische Bewertung von Biotreibstoffen. BfE/BAFU/BLW, Bern, Switzerland.

4 Bacterial Integrons

Shilpi Srivastava

CONTENTS

4.1 INTRODUCTION

Antibiotics are compounds that are used to kill disease-causing microbes or to limit their reproduction. These potent miracle drugs have been extensively used since their discovery and have saved millions of lives. Antibiotics have been extensively used for human medicine, in animal farming for disease therapy, and as feed additives for promoting growth and reducing the incidence of diseases. Therapeutic antibiotics are used in the treatment and control of a wide range of infections across the animal kingdom. Antibiotics have even been used in plant diseases where bacteria cause serious loss in crop productivity. Antibiotics induce mortality due to selective toxicity and disruption of microbial structures or processes that are nonexistent in human cells. Their production is driven by theories of *antibiosis*, that is, a human leveraging of substances that microbes create in mutually antagonistic battles for space and resources. Table 4.1 depicts the major antibiotics used and their resistance mechanisms.

The antibiotic industry is highly profitable, with annual sales of about USD 40 billion worldwide, encompassing both rich and poor countries and spreading across all levels of income. Antibiotics have been one of the most successful forms of chemotherapy in the history of medical science. They have saved countless lives and enabled the development of modern medicine over the past several decades.

The discovery of penicillin is attributed to Scottish Nobel Laureate Alexander Fleming, who demonstrated that *Penicillium rubens*, when grown in an appropriate substrate, exudes a substance having antimicrobial properties. Fleming called the substance as penicillin. This serendipitous observation initiated the

TABLE 4.1

Some Commonly Used Antibiotics and Their Modes of Resistance

Class of Antibiotic	Example	Modes of Resistance
Aminoglycosides	Gentamicin, streptomycin, spectinomycin	Acetylation, altered target, efflux, nucleotidylation, phosphorylation
β-Lactams	Ampicillin, aztreonam, cephamycin, meropenem	Altered target, efflux, hydrolysis
Cationic peptides	Colistin	Altered target, efflux
Glycopeptides	Vancomycin, teicoplanin	Reprogramming of peptidoglycan biosynthesis
Macrolides	Azithromycin, erythromycin	Altered target, efflux, glycosylation, hydrolysis, phosphorylation
Phenicols	Chloramphenicol	Acetylation, altered target, efflux
Quinolones	Ciprofloxacin	Acetylation, altered target, efflux
Rifamycins	Rifampin	Adenosine diphosphate-ribosylation, altered target, efflux
Sulfonamides	Sulfamethoxazole	Altered target, efflux
Tetracyclines	Minocycline, tigecycline	Altered target, efflux, monooxygenation

era of antibiotic discovery and is continuing even today. However, penicillin was developed as a drug many years later by Norman Heatley, Ernst Chain, and Howard Florey in wartime England. A breakthrough by the German Biochemist Gerhard Domagk at the Bayer Laboratories in 1935 ushered the era of antibacterials. Domagk discovered and developed the first sulfonamide-based synthetic drug popularly known as Prontosil. Prontosil was the first commercially available antibacterial drug. Fleming, after discovering penicillin, tried hard to resolve the persisting problems with purification and stability of the active substance, but due to lack of support from chemists, he finally abandoned the idea in 1940. However, an Oxford team led by Howard Florey and Ernest Chain described the purification of penicillin, which was sufficient enough for clinical testing. This development eventually led to the mass production and distribution of penicillin in 1945. It effectively treated bacterial infections in previously untreatable diseases, and its greater efficacy and relatively fewer side effects than therapeutic agents made it appear as a *miracle drug*. The use of antibiotics to treat human infections started with sulfonamides and was followed by the aminoglycoside streptomycin and streptothricin. Streptomycin was introduced in 1944 for the treatment of tuberculosis. Thus, the *Golden Age* of antibiotics began, which started in the 1940s and extended till the 1990s. During this period, almost all antibacterial spectra with different generations such as β-lactams, chloramphenicol, aminoglycosides, tetracyclines, glycopeptides, macrolides, streptogramins, and quinolones were developed and introduced in clinical practice. The discovery of antibiotics is of great significance, since it has greatly improved the quality of human life in the twentieth and the twenty-first centuries.

4.2 THE ANTIBIOTIC INDUSTRY

The popularity of antibiotics has given rise to a huge industry that caters to the needs of this enlarging industry. Pharmaceutical companies discover, develop, and market new antibacterials that have been used for more than 50 years to improve both human and animal health. The use of antibiotics is not just limited to the pharmaceutical industry but also extends to the agricultural sector. Antibiotics are used in the husbandry of livestock in the treatment or prophylaxis of infection, to promote growth, and to improve feed efficiency in farm animals. Antibiotics have been extensively used in to reduce the suffering of diseased animals and to control the spread of the illness to healthy animals. Antibiotics, along with antifungals and other drugs, are used by veterinarians and livestock owners to increase the growth rates of poultry, livestock, and other farm animals. In human medicine, antibiotics have changed the way medicine is practiced, by being a major player in the diagnosis and treatment of infections. Antibiotics supplement the body's natural defense mechanism against microbial infections and are in high demand than any other drugs. This provides ample evidence of their critical role in the world economy and public health.

Antibiotics are in great demand ever since their discovery. In 2009, the antibiotics market generated sales of USD 42 billion globally, representing 46% of sales of anti-infective agents and 5% of the global pharmaceutical market. The sales of diverse classes of antibiotics are depicted in Figure 4.1.

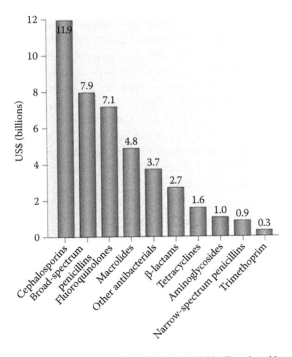

FIGURE 4.1 Sales of different classes of antibiotics in 2009. (Reprinted by permission from Macmillan Publishers Ltd. *Nat. Rev. Drug Disc.*, Hamad, B., 2010, copyright 2010.)

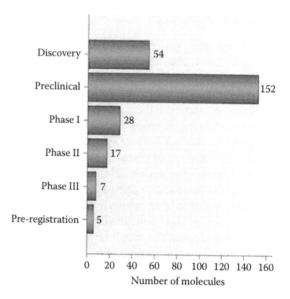

FIGURE 4.2 Status of antibiotic research and development activity at each development stage. (Reprinted by permission from Macmillan Publishers Ltd. *Nat. Rev. Drug Disc.*, Hamad, B., 2010, copyright 2010.)

Since 2005, this market has grown at an annual rate of 6.6% until 2011. The demand for antibiotics is expected to reach USD 44.68 billion by the year 2016. By 2010, more than 150 antibiotic molecules were in preclinical development, but only 17 molecules were in Phase II and 7 molecules were in Phase III trials (Figure 4.2).

Bayer HealthCare AG, Bristol-Myers Squibb Co., Abbott Laboratories, Roche, Astellas Pharma Inc., MiddleBrook Pharmaceuticals, Daiichi Sankyo Company Ltd., Kyorin Pharmaceutical Co. Ltd., Cubist Pharmaceuticals Inc., Takeda Pharmaceutical Company Ltd., Eli Lilly and Co., GlaxoSmithKline plc, Johnson and Johnson, Pfizer Inc., LG Life Sciences Limited Inc., Novartis AG, Sanofi-Aventis SA, and Toyama Chemical Co. Ltd. are some of the major companies involved in the development and production of antibiotics.

4.3 ANTIBIOTIC RESISTANCE IN BACTERIA

Soon after the introduction of antibiotics into clinical practice in the 1940s, they were a huge success, and this led to a general perception that infectious diseases would become a problem of the past and would be eventually eliminated from the world. In contrast, the use of antibiotics promoted the development of antibiotic-resistant bacteria. The large-scale production and use of antibiotics in various sectors such as clinical and veterinary medicine, horticulture, agriculture, aquaculture, and other human activities resulted in a massive explosion of antibiotic-resistant phenotypes in both human and animal pathogens. Antibiotic resistance originates when the microbe modifies in a way that reduces or eliminates the effectiveness of drugs that were designed to cure infections. The World Health Organization (WHO) defines antimalarial drug resistance as the ability of a strain to survive and/ or multiply despite administration and absorption of a drug given in doses equal to or higher than those usually recommended but within the tolerance limit of the patient. Microbes can also develop cross-resistance to drugs belonging to the same chemical family or against those that share similar modes of action. So, the bacteria survive and continue to multiply even after administration of the drug, causing more damage than before. Resistance to antibiotics existed even before the use of antibiotics. However, this intrinsic form of resistance is not a major source of concern in comparison with the emergence of a vast majority of drug-resistant organisms that originate by genetic modifications or by transfer of genetic material and subsequent selection processes. The administration of penicillin to patients in 1941 led to the emergence of penicillin-resistant bacteria within a year in 1942. Resistance to antibiotics is now one of our most serious health threats. Pathological conditions resulting from resistant bacteria have slowly increased, with some pathogens becoming resistant to multiple types or classes of antibiotics. *Shigella* strains, observed in Japan in the 1950s, were the first documented multidrug-resistant (MDR) isolates that contained plasmids, then called R factors, which could transfer the resistances to antibiotic-sensitive cells, thereby spreading the resistance genes to new organisms. Bacteria have become resistant to antimicrobials through a number of mechanisms, discussed as follows:

1. The access of antimicrobials to target sites is restricted through permeability changes in the bacterial cell wall.
2. Modification of the antibiotics.
3. Active efflux of the antibiotic from the microbial cell.
4. Degradation of the antibiotic.
5. Development of alternative metabolic pathways to those inhibited by the drug.
6. Modification of antibiotic targets.
7. Overproduction of the target enzyme.

Some of the above-mentioned major mechanisms of disease resistance have been depicted in Figure 4.3. The genetic exchange occurs among bacteria of different taxonomic and ecological groups for genes for resistance traits. It is usually accomplished through mobile genetic elements such as transposons, plasmids, naked DNA, and bacteriophages (Figure 4.4).

Many of the pathogens associated with human diseases have evolved into MDR forms after the rampant use of antibiotics. *Shigella* strains were the first MDR isolates documented in Japan in the 1950s. Some of the most problematic MDR organisms that are encountered currently include *Pseudomonas aeruginosa* (another microbe of soil origin), *Acinetobacter baumannii*, *Escherichia coli*, and *Klebsiella pneumoniae*, bearing extended-spectrum β-lactamases (ESBL), vancomycin-resistant enterococci (VRE), methicillin-resistant *Staphylococcus aureus* (MRSA), vancomycin-resistant

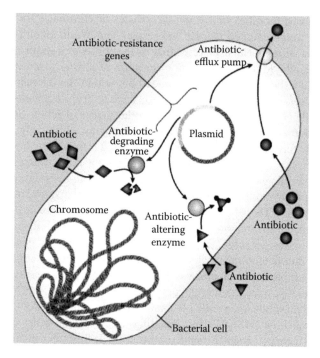

FIGURE 4.3 Biological mechanisms of resistance against antibiotics in bacteria. (Reprinted by permission from Macmillan Publishers Ltd. *Nat. Med.*, Levy, S.B. and Marshall, B., 2004, copyright 2004.)

S. aureus, and extensively drug-resistant (XDR) *Mycobacterium tuberculosis*. The combination of resistance with virulence acts as a potentially deadly outcome, as evident in the epidemic outbreak of *E. coli* 0104:H4 in Europe. There have been reports of exceptional virulence capabilities of MDR *S. aureus*. Table 4.2 provides information on the resistance of bacteria to specific class of antibiotics. Multidrug resistance is a worldwide problem that can indiscriminately affect members of all socioeconomic classes across countries.

The migration of a strain of MDR *Streptococcus pneumoniae* from Spain to the United States, the United Kingdom, and South Africa has been documented. In many countries, antibiotics are considered *ordinary* drugs and are prescribed freely by different physicians, both in the community and in hospitals. The situation is even more alarming in the developing countries, where the antibiotics are available over the counter. Whenever resistance becomes a clinical problem in the developing countries, these countries may have no substitutes available, since they often do not have access to expensive drugs. China is one of the countries severely affected by antibiotic use, where the annual prescription of antibiotics (both clinical and veterinary) is approaching 140 gram per person, which is about 10 times higher than that prescribed in the United Kingdom. Many scientific gatherings, task forces, workshops, and international meetings have pointed out toward this

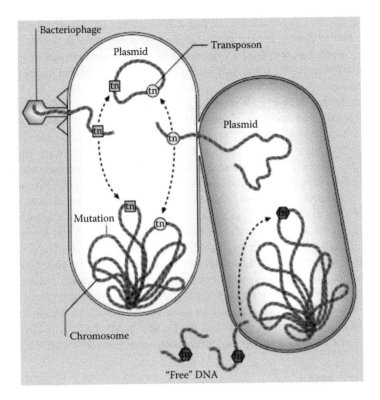

FIGURE 4.4 Genetics of spread of drug resistance in bacteria. (Reprinted by permission from Macmillan Publishers Ltd. *Nat. Med.*, Levy, S.B. and Marshall, B., 2004, copyright 2004.)

TABLE 4.2
Different Bacterial Species Resistant to Antibiotics in Nature

Bacteria	Antibiotics
Acinetobacter spp.	β-lactams, fluoroquinolones, aminoglycosides
Escherichia coli	Third-generation cephalosporins, including resistance conferred by extended-spectrum β-lactamases, and fluoroquinolones
Helicobacter pylori	Tetracycline
Klebsiella pneumoniae	Third-generation cephalosporins, including resistance conferred by extended-spectrum β-lactamases, and carbapenems
Megasphaera elsdenii	Tetracycline
Neisseria gonorrhoeae	Decreased susceptibility to third-generation cephalosporins
Pseudomonas aeruginosa	β-lactams, fluoroquinolones, aminoglycosides
Shigella spp.	Resistance to fluoroquinolones
Staphylococcus aureus	Resistance to β-lactam antibacterial drugs (methicillin, methicillin-resistant *S. aureus* [MRSA])
Streptococcus pneumoniae	Resistance or nonsusceptibility to penicillin (or both), telithromycin

threat over the last two decades but had a limited impact due to a lack of coordination. The antibiotic consumption has been reduced only in a few developed countries, which has resulted in a decrease in resistance. However, despite these efforts, resistance among some bacterial species is on the rise and has been gradually increasing.

The emergence of highly resistant strains of gram-negatives, viz. *P. aeruginosa, K. pneumoniae, Enterobacter cloacae, A. baumannii,* and *E. coli,* presents the greatest threat to human life and modern medicine. These bacteria cause infections, ranging from urinary tract infections and bloodstream infections to life-threatening pneumonias. The cell wall of these bacteria is difficult to penetrate, and even on entry, the bacteria pump the antibiotics out, again leading to less effectiveness of the antibiotics. The problem is more critical in gram-negative bacteria, where the MDR strains are on the rise and few antibiotics are available that can currently combat them. Many of the antibiotic products currently in the developmental pipeline would be active against some of the above-listed pathogens. Only one-third (approximately 16 drugs) show significant activity against MDR gram-negative species. Of these, only three *potential breakthrough antibiotics* would offer results against the vast majority of the resistant bacteria known today. A careful assessment points out to the hard fact that the antibiotics currently in development—some of which will hit the markets 10 to 15 years from now—will not be able to fill the gaps in clinical need that already exist, and these gaps are more likely to increase as resistance spreads.

4.4   INTEGRONS

Integrons are versatile gene acquisition systems commonly found in bacterial genomes that allow efficient capture and expression of exogenous genes. These are assembly platforms that incorporate exogenous open reading frames (ORFs) through site-specific recombination (a form of sexual recombination), ensure their correct expression, and convert them into functional genes. Integrons are known to occur in varied types of environments, are able to move between species and lineages over evolutionary time frames, and have access to a vast pool of novel genes, whose functions are yet to be determined.

Integrons consists of three main components (Figure 4.5):

1. The gene encoding a tyrosine recombinase (integrase, encoded by the *intI* gene), which is a prerequisite for site-specific recombination within an integron.
2. The primary recombination site (*attI*) that is located in an adjacent region and is recognized by the integrin integrase.
3. The promoter (Pc), which is required for proper transcription and expression of gene cassettes present in the integron. The location of the promoter is to the upstream of the integration site.

An integron usually contains one or more gene cassettes that have been incorporated into it. The *attC* site is generally associated with a single ORF expressed from the Pc promoter in a structure termed a gene cassette, which forms the mobile component

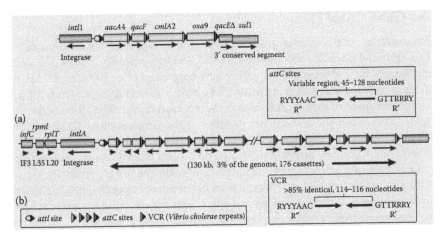

FIGURE 4.5 Mobile integrons and superintegrons. Structural comparison of a *classical* mobile integron and the superintegron from *Vibrio cholerae* strain N16961. (a) A schematic representation of the class 1 integron In40. The various resistance-gene cassettes carry different *attC* sites. (b) Schematic representation of the chromosomal superintegron in *V. cholerae*; the open reading frames are separated by highly homologous sequences, the *V. cholerae* repeats (VCRs). infC encodes translation initiation factor IF3; rpmI and rplT encode ribosomal proteins L35 and L20, respectively. (Reprinted by permission from Macmillan Publishers Ltd. *Nat. Rev. Microbiol.*, Mazel, D., 2006, copyright 2006.)

of the system. All integron-inserted gene cassettes identified so far generally contain a single gene and an imperfect inverted repeat *attC* site (initially called a 59-base element) at the 3′ end of the gene. The *attC* sites are a diverse family of nucleotide sequences, varying in length from 57 to 141 bp. They form the recognition sites for the site-specific integrase. The nucleotide sequence similarities of the *attC* sites are primarily restricted to the edges that contain conserved sequences known as the R″ sequence (RYYYAAC, where R = purine and Y = pyrimidine) and the R′ sequence (GTTRRRY, where the point of recombination is between the G and T bases). Circular gene cassettes are integrated by a reversible site-specific recombination between *attI* and *attC*, mediated by the integron integrases. New gene cassettes are acquired by the integrons through recombination between the *attI* site of the integron and the *attC* of a circular cassette.

Recombination in integrons is carried out by the integron integrases, a distinct group of tyrosine (Y)-recombinases closely related to XerCD recombinases, which rearrange DNA duplexes by means of conservative site-specific recombination reactions. These Y-recombinases are most widespread among prokaryotes, where they play fundamental roles such as the integration/excision of viral genomes, the alternation of gene expression, and the resolution of deleterious chromosome dimers arising during replication. Y-recombinases show a conserved fold of the catalytic domain, including a three-dimensional clustering of highly conserved RKHRH residues, and recognize specific DNA sites comprising two inverted binding domains separated by a 6–8 bp spacer.

4.5 GENE CASSETTES

Gene cassettes form a diverse group of small mobile elements that contains only a single gene and a specific recombination site. Cassettes generally consist of a single gene flanked by a short sequence called a 59-base element or 59-be of 57–141 bp, located downstream of the gene. Most 59-be have a central axis of symmetry and also contain inverted repeat sequences in their central region. The 59-be is the specific recombination site and confers mobility to the gene cassettes. Cassettes generally do not include a promoter and are transcribed from a promoter located in the integron. Cassettes carrying their own promoters are expressed independently of the constitutive promoter (PC), and their initiation codons are normally located pretty close to one boundary. Thus, the multicomponent cassette–integron system can be considered a natural cloning and expression system. Most cassettes are possibly expressed from a common promoter situated in the conserved 5′ segment of the integron, which promotes coexpression of the inserted gene cassettes from a single promoter, and selection for one resistance determinant often coselects for the maintenance of the entire array. Integron gene cassettes are abundant in environmental samples and can be recovered from diverse environments such as soils, sediments, hot springs, estuaries, seawater, marine sediment, deep sea vents, plant surfaces, symbionts of eukaryotes, and even complex organizations such as biofilms. Environmental gene cassettes are bestowed with different functions such as virulence, secondary metabolism, maintenance of plasmids, and surface properties. Gene cassettes are gigantic reservoirs of genomic novelty, with distinctive populations of gene cassettes found in diverse environments, having minute overlapping in composition across environments. No verifiable homologues of about 65% of the gene cassettes and their encoded polypeptides have been documented in DNA or protein databases.

About 15% of the gene cassettes exhibit homology to conserved hypothetical proteins, whereas the remaining have sufficient homology to characterized proteins, which can be used to predict their function. Though numerous polypeptides encoded by the gene cassettes are predicted to form novel protein folds and in a way comprise a toolbox of flexible molecular components for assembling new quaternary structures, the noncoding cassettes can make up a considerable proportion of arrays in some microbes.

Gene cassettes harbor genes having antibiotic resistance, and their accumulation in integrons has created multiresistance integrons (MRIs), due to which the antibiotic resistance problem has been on the rise. The unique *attC* sites have been a major player in about 130 antibiotic-resistance genes and confer resistance to the following classes of antibiotics: all known β-lactams, all aminoglycosides, rifampin, erythromycin, chloramphenicol, streptothricin, trimethoprim, quinolones, fosfomycin, lincomycin, and antiseptics of the quaternary ammonium-compound family. Integrons having up to eight resistance gene cassettes have been reported in multiple-resistant clinical isolates. Cassettes are usually known either by the name of the gene encoded, or in the case of cassettes containing ORFs with unknown functions, the ORFs have been assigned letters in the order of their identification and the cassettes are identified by these names. It is assumed that many of the nonexpressed cassettes usually survive in the absence of selective pressure and provide a genetic basis for the evolution and spread of novel functions.

4.6   EVOLUTIONARY HISTORY

Integrons are ancient entities that have been intricately associated with the evolution of bacterial genomes since millions of years. The fact that clustering of the respective superintegron integrase genes takes place according to species and adheres to the line of descent indicates that they are ancient structures. Integrons are found in a wide range of environments, including soils, riverine sediments, marine sediments, deep-sea sediments, plant surfaces, aquatic biofilms, and even hot springs with more than 15% of genome-sequenced bacteria.

4.7   TYPES OF INTEGRONS

Initially, integrons were divided into two groups:

1. *Mobile integrons*: These have different *attC* sites but only a few cassettes encoding antibiotic resistance. Mobile integrons have gained their mobility by association with plasmids or transposons.
2. *Superintegrons*: These were known to have homogenous *attC* sites and hundreds of cassettes. They were situated on chromosomes.

However, it was later discovered that there is a series of integron structures between these two extremes.

Hundreds of different integron families have been discovered, which can be differentiated on the basis of the relative homology of *intI*, the gene encoding the integron integrase. Based on the phylogeny of their respective integrase genes, the integrons fall into three major groups:

1. Group present in proteobacteria from freshwater and soil environments
2. Group found in gamma proteobacteria from marine environments
3. Integrons whose integrase genes are in the reverse orientation to those listed earlier

However, the most accepted classification is based on the dissimilarities and divergence in the sequences of *intI*. Based on these criteria, integrons have been divided into four types, all of which are physically linked to mobile DNA elements such as transposons, insertion sequences (ISs), and conjugative plasmids that are responsible for the intra- and interspecies transfer of the genetic material.

1. *Class I integrons*: These are the most ubiquitous, are most commonly reported among clinical bacteria, and are found in approximately 9% of the sequenced bacterial genomes. They have spread into more than 70 bacterial

species of medical importance, and the conserved motifs of this class suggest that a single, recent ancestor gave rise to all current variants. Class 1 integron has been elaborately reported in several gram-negative microorganisms and a few gram-positive bacteria. Class 1 integrons have been reported in *Acinetobacter, Aerococcus, Aeromonas, Alcaligenes, Brevibacterium, Burkholderia, Campylobacter, Citrobacter, Corynebacterium, Enterobacter, Enterococcus, Escherichia, Klebsiella, Mycobacterium, Pseudomonas, Salmonella, Serratia, Shigella, Staphylococcus, Stenotrophomonas, Streptococcus,* and *Vibrio.* A gene cassette has also been discovered in *Enterococcus faecalis.* Class 1 integron does not exhibit self-movability but seem to be directly linked with Tn402-like transposons and associated with Tn3 transposon family (Tn21 or Tn1696). Class 1 integrons have also been found to be associated with diverse IS elements such as IS26, IS1999, IS2000, and IS6100, with IS6100 most frequently found at the 3′ end of integrons. These are the most common and widespread, especially in clinical settings, and have been found in bacterial isolates obtained from cattle, swine, chickens, fish, pet dogs, zoo animals, and even apple orchard. Class 1 integrons are a major contributor to the evolution and dissemination of antibiotic resistance with the gene cassettes having the ability to conduct horizontal gene transfer among microorganisms.

2. *Class 2 integrons*: Class 2 integrons are frequently found to be associated with the Tn7 transposon family (Tn7 and its derivatives, such as Tn1825, Tn1826, and Tn4132), carrying both the promoter Pc and the recombination site *attI2*. These contain an array of gene cassettes, which comprises streptothricin acetyltransferase (*sat1*), dihydrofolate reductase (*dfrA1*), and aminoglycoside adenyltransferase (*aadA1*) that confer resistance to streptothricin, trimethoprim, and streptomycin/spectinomycin, respectively. Most of the gene cassette arrays of class 2 integrons are conserved and show integrase inactivation, due to which there is a lack of dynamic recombination. The mobility of class 2 integron is carried out by preferential insertion into a unique site within the bacterial chromosome, a step mediated by five tns genes, viz. *tnsA, tnsB, tnsC, tnsD,* and *tnsE.* Class 2 integrons have been reported in *E. coli, Shigella flexneri, P. aeruginosa, A. baumannii, Proteus vulgaris, Proteus mirabilis,* Enterobacteriaceae, and *Salmonella.*

3. *Class 3 integrons*: Class 3 introns were first reported in *Serratia marcescens* in the year 1993 and were later identified in *K. pneumoniae, Acinetobacter* spp., *Alcaligenes, Citrobacter freundii, E. coli, P. aeruginosa, Pseudomonas putida,* and *Salmonella* spp. The structure of class 3 integron is similar structure to that of class 2 introns, since both *IntI1* and *IntI3* are part of the proteobacteria group.

4. *Class 4 integrons*: Class 4 integrons were first detected in *Vibrio* sp. and are thought to exist before the antibiotic era. Class 4 integrons are found in microbes such as *Vibrionaceae, Shewanella, Xanthomonas,* Pseudomonad, and other proteobacteria and carry gene cassettes that impart resistance to several antibiotics such as chloramphenicol and fosfomycin. Class 4

integrons harbor a large array of gene cassettes that provides the bacteria with different adaptations, with extension beyond antibiotic resistance and pathogenicity. This distinctive class of integron is distinguished from other classes by two key features: incorporation of hundreds of cassettes and high homology between the *attC* sites of those gathered cassettes.

4.8   SUPERINTEGRONS

The terms *superintegron*, coined by Mazel et al. (1998), and *mega-integron*, coined by Clark et al. (1997), refer to all the integrons found in a bacterial chromosome or full bacterial genome, including large arrays of cassettes. The initial use of the terms *superintegron* and *mega-integron* by both the research groups referred to an integron having a very large array of gene cassettes incorporated within it, found in the small chromosome of *Vibrio cholerae*, a facultative anaerobe. Vaisvila et al. (2001) defined integron as a structure containing only 30 cassettes in *Pseudomonas alcaligenes* strain that carried three different integrons. The *V. cholerae* superintegron is 126-kb long and contains about 179 cassettes, since 179 *V. cholerae* repetitive (VCR) sequence copies can be identified. It is believed that some of the cassettes in *V. cholerae* have been acquired through lateral transfer from the superintegron cassette pool of another bacterial species.

4.9   ROLE IN THERAPEUTICS

Integrons play a major role in the development of antibiotic resistance among a range of pathogenic organisms and have contributed to the adaption and evolution of bacteria. Integrons have increased enormously in abundance, interacted with other DNAs, gathered a large number of resistance genes from the environment, and have given rise to new complex mobile elements that carry resistance against a range of stresses such as antibiotics, disinfectants, and heavy metals.

There has been a quick and widespread appearance of multiple mutations that confer multidrug resistance in unrelated bacteria. This unique phenomenon cannot be attributed to mutation alone and requires the movement of resistant genes through transferable genetic elements. However, detailed studies of the mobile genetic elements with different antibiotic-resistance spectra pointed out to the involvement of a new type of genetic element termed an integron. The mobile DNA elements associated with integrons facilitate easy transfer of the resistance genes that have been accumulated by integrons cutting across phylogenetic boundaries, which points toward the positive impact of integrons on bacterial evolution. The functions encoded by the MRIs and superintegron cassettes have suggested that both code for proteins related to simple adaptive functions, and their presence provides the pathogenic bacteria with a selective advantage.

Integrons have given an impetus to the spread of antibiotic resistance, especially in gram-negative bacteria (Table 4.3).

Gene cassettes coding for antibiotic resistance, their accumulation in the integrons, and creation of MDR integrons have contributed to the current resistance problem to a great extent. Antibiotic-resistance integrons usually exhibit mobility,

TABLE 4.3
Integron-Associated Antibiotic Resistance in Bacteria

Bacteria	Gram +/−	Antibiotic	Reference
Acinetobacter baumannii	−	β-lactams	Riccio et al. (2000)
Campylobacter jejuni	−	Trimethoprim	Gibreel and Skold (2000)
		Tobramycin-gentamicin	Lee et al. (2002)
Corynebacterium glutamicum	+	Streptomycin, spectinomycin, tetracycline	Tauch et al. (2002)
Enterobacter cloacae	−	β-lactams	Navon-Venezia et al. (2008)
Enterococcus faecalis	+	Streptomycin/ spectinomycin	Clark et al. (1999)
Escherichia coli	−	β-lactams	Navon-Venezia et al. (2008)
Klebsiella pneumoniae	−	β-lactams	Navon-Venezia et al. (2008)
Pasteurella aerogenes	−	Trimethoprim	Schwarz and Kehrenberg (2011)
Pseudomonas aeruginosa	−	β-lactams	Mavroidi et al. (2001)
Salmonella enterica	−	β-lactams	Di Conza et al. (2002)
Serratia marcescens	−	β-lactams	Osano et al. (1994)
Staphylococcus aureus	+	Methicillin	Xu et al. (2008, 2011)
Vibrio cholera	−	β-lactams	Dalsgaard et al. (2000)
		Chloramphenicol, tetracycline, streptomycin	Iwanaga et al. (2004)

with their cassettes arrays being short and encoding antibiotic resistance. The location of a cassette in the integron (both order and distance) has a great bearing on the level of antibiotic resistance.

4.10   CONCLUSION

The use of antimicrobial agents as therapeutic agents has contributed substantially to human's increased life span by reducing the morbidity and mortality of humans. There has been a rapid emergence of bacterial resistance, especially multiple antibiotic resistance, which is a direct fallout of the increased use of antibiotics. Integrons as mobile genetic elements have played a major role in the horizontal transfer of antibiotic resistance among bacteria, which has been well researched in recent decades. The discovery of larger integron structures known as superintegrons has further increased our understanding of their significant role in bacterial genome evolution. Integrons and superintegrons are known to play a major role in antibiotic resistance among clinically important microorganisms and contribute to the adaptation and evolution of bacteria. In the coming years, integron research would focus on elucidating the mechanisms underpinning the recombination process, the genesis of new gene cassettes, and the mechanism of exchange of gene cassette in bacterial populations.

BIBLIOGRAPHY

Alekshun M.N. and Levy S.B. 2000. Bacterial drug resistance: Response to survival threats. In: Storz G. and Hengge-Aronis R. (Eds.) *Bacterial Stress Responses*. ASM Press, Washington, DC, pp. 323–366.

Allen H.K., Donato J., Wang H.H., Cloud-Hansen K.A., Davies J. and Handelsman J. 2010. Call of the wild: Antibiotic resistance genes in natural environments. *Nat. Rev. Microbiol.* 8: 251–259.

Aminov R.I. 2010. A brief history of the antibiotic era: Lessons learned and challenges for the future. *Front. Microbiol.* 1: 134.

Aminov R.I. and Mackie R.I. 2007. Evolution and ecology of antibiotic resistance genes. *FEMS Microbiol. Lett.* 271: 147–161.

Bax R.P. 1997. Antibiotic resistance: A view from the pharmaceutical industry. *Clin. Infect. Dis.* 24 (Suppl 1): S151–S153.

Boucher Y.L.M., Koenig J.E. and Stokes H.W. 2007. Integrons: Mobilizable platforms that promote genetic diversity in bacteria. *Trends Microbiol.* 15: 301–309.

Buchholz U., Bernard H., Werber D., Böhmer M.M., Remschmidt C., Wilking H., Deleré Y. et al. 2011. German outbreak of *Escherichia coli* O104: H4 associated with sprouts. *N. Engl. J. Med.* 365: 1763–1770.

Carlet J., Collignon P., Goldmann D., Goosens H., Gyssens I.C., Harbarth S., Jarlier V. et al. 2011. Society's failure to protect a precious resource: Antibiotics. *Lancet* 378: 369–371.

Carlet J., Jarlier V., Harbarth S., Voss A., Goossens H. and Pittet D. 2012. Ready for a world without antibiotics? The pensières antibiotic resistance call to action. *Antimicrob. Resist. Infect. Control* 1: 11.

Clark C.A., Purins L., Keawrakon P. and Manning P.A. 1997. VCR repetitive sequence elements in the *Vibrio cholerae* chromosome constitute a mega-integron. *Mol. Microbiol.* 26: 1137–1138.

Clark C.A., Purins L., Kaewrakon P., Focareta T. and Manning P.A. 2000. The *Vibrio cholerae* O1 chromosomal integron. *Microbiology* 146: 2605–2612.

Clark N.C., Olsvik O., Swenson J.M., Spiegel C.A. and Tenover F.C. 1999. Detection of a streptomycin/spectinomycin adenylyltransferase gene (aadA) in *Enterococcus faecalis*. *Antimicrob. Agents Chemother.* 43: 157–160.

Collis C.M. and Hall R.M. 1992. Gene cassettes from the insert region of integrons are excised as covalently closed circles. *Mol. Microbiol.* 6: 2875–2885.

Collis C.M. and Hall R.M. 1995. Expression of antibiotic resistance genes in the integrated cassettes of integrons. *Antimicrob. Agents Chemother.* 39: 155–162.

Collis C.M., Grammaticopoulos G., Briton J., Stokes H.W. and Hall R.M. 1993. Site-specific insertion of gene cassettes into integrons. *Mol. Microbiol.* 9: 41–52.

Corpet D.E. 1988. Antibiotic resistance from food. *N. Engl. J. Med.* 318: 1206–1207.

Courvalin P. 1994. Transfer of antibiotic resistance genes between gram-positive and gram-negative bacteria. *Antimicrob. Agents Chemother.* 38: 1447–1451.

Courvalin P. 2008. Predictable and unpredictable evolution of antibiotic resistance. *J. Intern. Med.* 264: 4–16.

Dalsgaard A., Forslund A., Serichantalergs O. and Sandvang D. 2000. Distribution and content of class 1 integrons in different *Vibrio cholerae* O-serotype strains isolated in Thailand. *Antimicrob. Agents Chemother.* 44: 1315–1321.

Davies J. and Davies D. 2010. Origins and evolution of antibiotic resistance. *Microbiol. Mol. Biol. Rev.* 74: 417–433.

Davies J.E. 1994. Inactivation of antibiotics and the dissemination of resistance genes. *Science* 264: 375–382.

Davies J.E. 1997. Origins, acquisition and dissemination of antibiotic resistance determinants. *Ciba Found. Symp.* 207: 15–27.

Deng Y., Bao X., Ji L., Chen L., Liu J., Miao J., Chen D., Bian H., Li Y. and Yu G. 2015. Resistance integrons: Class 1, 2 and 3 integrons. *Ann. Clin. Microbiol. Antimicrob.* 14: 45.

Devasahayam G., Scheld W.M. and Hoffman P.S. 2010. Newer antibacterial drugs for a new century. *Expert Opin. Investig. Drugs* 19: 215–234.

Di Conza J., Ayala J.A., Power P., Mollerach M. and Gutkind G. 2002. Novel class 1 integron (InS21) carrying *bla*CTX-M-2 in *Salmonella enterica* serovar infantis. *Antimicrob. Agents Chemother.* 46: 2257–2261.

Domingues S., da Silva G.J. and Nielsen K.M. 2012. Integrons: Vehicles and pathways for horizontal dissemination in bacteria. *Mob. Genet. Elements* 2: 211–223.

Escudero J.A., Loot C., Parissi V., Nivina A., Bouchier C. and Mazel D. 2016. Unmasking the ancestral activity of integron integrases reveals a smooth evolutionary transition during functional innovation. *Nat. Commun.* 7: 10937.

Fluit A.C. and Schmitz F.J. 1999. Class 1 integrons, gene cassettes, mobility, and epidemiology. *Eur. J. Clin. Microbiol. Infect. Dis.* 18: 761–770.

Fluit A.C. and Schmitz F.J. 2004. Resistance integrons and super-integrons. *Clin. Microbiol. Infect.* 10: 272–288.

Gibreel A. and Skold O. 2000. An integron cassette carrying dfr1 with 90-bp repeat sequences located on the chromosome of trimethoprim-resistant isolates of *Campylobacter jejuni*. *Microb. Drug Resist.* 6: 91–98.

Gillings M.R. 2017. Class 1 integrons as invasive species. *Curr. Opin. Microbiol.* 38: 10–15.

Goossens H. 2009. Antibiotic consumption and link to resistance. *Clin. Microbiol. Infect.* 15: 12–15.

Hall R.M. and Collis C.M. 1995. Mobile gene cassettes and integrons: Capture and spread of genes by site-specific recombination. *Mol. Microbiol.* 15: 593–600.

Hall R.M. and Collis C.M. 1998. Antibiotic resistance in gram negative bacteria: The role of gene cassettes and integrons. *Drug Resist. Updates* 1: 109–119.

Hall R.M., Collis C.M., Kim M.J., Partridge S.R., Recchia G.D. and Stokes H.W. 1999. Mobile gene cassettes and integrons in evolution. *Ann. NY Acad. Sci.* 870: 68–80.

Hall R.M. and Stokes H.W. 1993. Integrons: Novel DNA elements which capture genes by site-specific recombination. *Genetica* 90: 115–132.

Hamad B. 2010. The antibiotics market. *Nat. Rev. Drug Disc.* 9: 675–676.

Hanau-BerAot B., Podglajen I., Casin I. and Collatz E. 2002. An intrinsic control element for translational initiation in class 1 integrons. *Mol. Microbiol.* 44: 119–130.

Heuer H., Schmitt H. and Smalla K. 2011. Antibiotic resistance gene spread due to manure application on agricultural fields. *Curr. Opin. Microbiol.* 14: 236–243.

Hughes J.M. 2011. Preserving the lifesaving power of antimicrobial agents. *JAMA* 305: 1027–1028.

Iwanaga M., Toma C., Miyazato T., Insisiengmay S., Nakasone N. and Ehara M. 2004. Antibiotic resistance conferred by a class I integron and SXT Constin in *Vibrio cholerae* O1 strains isolated in Laos. *Antimicrob. Agents Chemother.* 48: 2364–2369.

Labbate M., Case R.J. and Stokes H.W. 2009. The integron/gene cassette system: An active player in bacterial adaptation. *Methods Mol. Biol.* 532: 103–125.

Landers T.F., Cohen B., Wittum T.E. and Larson E.L. 2012. A review of antibiotic use in food animals: Perspective, policy, and potential. *Public Health Rep.* 127: 4–22.

Lee M.D., Sanchez S., ZImmer M., Idris U., Berrang M.E. and Mcdermott P.F. 2002. Class 1 integron-associated tobramycin-gentamicin resistance in *Campylobacter jejuni* isolated from the broiler chicken house environment. *Antimicrob. Agents Chemother.* 46: 3660–3664.

Levin B.R., Lipsitch M., Perrot V., Schrag S., Antia R., Simonsen L., Walker N.M. and Stewart F.M. 1997. The population genetics of antibiotic resistance. *Clin. Infect. Dis.* 24 (Suppl 1): S9–S16.

Levy S.B. 1982. Microbial resistance to antibiotics. An evolving and persistent problem. *Lancet* 2: 83–88.

Levy S.B. 1984. Antibiotic resistant bacteria in food of man and animals. In: Woodbine, M. (Ed.) *Antimicrobials and Agriculture.* Butterworths, London, UK, pp. 525–531.

Levy S.B. 1992. Active efflux mechanisms for antimicrobial resistance. *Antimicrob. Agents Chemother.* 36: 695–703.

Levy S.B. 1998. The challenge of antibiotic resistance. *Sci. Am.* 278: 46–53.

Levy S.B. 2001. Antibiotic resistance: Consequences of inaction. *Clin. Infect. Dis.* 33 (Suppl. 3): S124–S129.

Levy S.B. 2002. *The Antibiotic Paradox: How the Misuse of Antibiotics Destroys Their Curative Powers.* Perseus Publishing, Cambridge, MA.

Levy S.B. and Marshall B. 2004. Antibacterial resistance worldwide: Causes, challenges and responses. *Nat. Med.* 10: S122–S129.

Ma L., Li A.D., Yin X.L. and Zhang T. 2017. The prevalence of integrons as the carrier of antibiotic resistance genes in natural and man-made environments. *Environ. Sci. Technol.* 51(10): 5721–5728. doi:10.1021/acs.est.6b05887.

Machado E., Teresa M.C., Canton R., Sousa J.C. and Peixe L. 2008. Antibiotic resistance integrons and extended-spectrum b-lactamases among Enterobacteriaceae isolates recovered from chickens and swine in Portugal. *J. Antimicrob. Chemother.* 62: 296–302.

MARAN. 2005. Monitoring of antimicrobial resistance and antibiotic usage in animals in the Netherlands in 2005. CIDC-Lelystad, Lelystad, the Netherlands.

MARAN. 2007. Monitoring of antimicrobial resistance and antibiotic usage in animals in the Netherlands in 2006/2007. CIDC-Lelystad, Lelystad, the Netherlands.

Mavroidi A., Tzelepi E., Tsakris A., Miriagou V., Sofianou D. and Tzouvelekis L.S. 2001. An integron-associated beta-lactamase (IBC-2) from *Pseudomonas aeruginosa* is a variant of the extended-spectrum beta-lactamase IBC-1. *J. Antimicrob. Chemother.* 48: 627–630.

Mazel D. 2006. Integrons: Agents of bacterial evolution. *Nat. Rev. Microbiol.* 4: 608–620.

Mazel D. and Davies J. 1999. Antibiotic resistance in microbes. *Cell. Mol. Life Sci.* 56: 742–754.

Mazel D., Dychinco B., Webb V.A. and Davies J. 1998. A distinctive class of integron in the *Vibrio cholerae* genome. *Science* 280: 605–608.

Messier N. and Roy P.H. 2001. Integron integrases possess a unique additional domain necessary for activity. *J. Bacteriol.* 183: 6699–6706.

Navon-Venezia S., Chmelnitsky I., Leavitt A. and Carmeli Y. 2008. Dissemination of the CTX-M-25 family betalactamases among *Klebsiella pneumoniae, Escherichia coli* and *Enterobacter cloacae* and identification of the novel enzyme CTX-M-41 in *Proteus mirabilis* in Israel. *J. Antimicrob. Chemother.* 62: 289–295.

Osano E., Arakawa Y., Wacharotayankun R., Ohta M., Horii T., Ito H., Yoshimura F. and Kato N. 1994. Molecular characterization of an enterobacterial metallo-lactamase found in a clinical isolate of *Serratia marcescens* that shows imipenem resistance. *Antimicrob. Agents Chemother.* 38: 71–78.

Partridge S.R., Tsafnat G., Coiera E. and Iredell J.R. 2009. Gene cassettes and cassette arrays in mobile resistance integrons. *FEMS Microbiol. Rev.* 33: 757–784.

Piddock L.J. 2012. The crisis of no new antibiotics—What is the way forward? *Lancet Infect. Dis.* 12: 249–253.

Poirel L., Carrër A., Pitout J.D. and Nordmann P. 2009. Integron mobilization unit as a source of mobility of antibiotic resistance genes. *Antimicrob. Agents Chemother.* 53: 2492–2498.

Recchia G.D. and Hall R.M. 1995. Gene cassettes: A new class of mobile element. *Microbiology* 141: 3015–3027.

Recchia G.D. and Hall R.M. 1997. Origins of the mobile gene cassettes found in integrons. *Trends Microbiol.* 5: 389–394.

Riccio M.L., Francheschini N., Boschi L., Caravelli B., Cornaglia G., Fontana R., Amicosante G. and Rossolini G.M. 2000. Characterization of the metallo-lactamase determinant of *Acinetobacter baumannii* AC-54/97 reveals the existence of *bla*IMP allelic variants carried by gene cassettes of different phylogeny. *Antimicrob. Agents Chemother.* 44: 1229–1235.

Rowe-Magnus D.A. and Mazel D. 1999. Resistance gene capture. *Curr. Opin. Microbiol.* 2: 483–488.

Rowe-Magnus D.A. and Mazel D. 2001. Integrons: Natural tools for bacterial genome evolution. *Curr. Opin. Microbiol.* 4: 565–569.

Rowe-Magnus D.A. and Mazel D. 2002. The role of integrons in antibiotic resistance gene capture. *Intern. J. Med. Microbiol.* 292: 115–125.

Rowe-Magnus D.A., Davies J. and Mazel D. 2002a. Impact of integrons and transposons on the evolution of resistance and virulence. *Curr. Topics Microbiol. Immunol.* 24: 167–188.

Rowe-Magnus D.A., Guerout A.-M. and Mazel D. 1999. Super-Integrons. *Res. Microbiol.* 150: 641–651.

Rowe-Magnus D.A., Guerout A.M. and Mazel D. 2002. Bacterial resistance evolution by recruitment of super-integron gene cassettes. *Mol. Microbiol.* 43: 1657–1669.

Rowe-Magnus D.A., Guerout A.-M. and Mazel D. 2002b. Bacterial resistance evolution by recruitment of super-integron gene cassettes. *Mol. Microbiol.* 43: 1657–1679.

Schwarz S. and Kehrenberg C. 2011. Trimethoprim resistance in a porcine *Pasteurella aerogenes* isolate is based on a dfrA1 gene cassette located in a partially truncated class 2 integron. *J. Antimicrob. Chemother.* 66: 450–452.

Sherratt D.J. and Wigley D.B. 1998. Conserved themes but novel activities in recombinases and topoisomerases. *Cell* 93: 149–152.

Shlaes D., Levy S. and Archer G. 1991. Antimicrobial resistance- new directions. *ASM News* 57: 455–458.

Singha P., Chanda D.D., Maurya A.P., Paul D., Chakravarty A. and Bhattacharjee A. 2016. Distribution of Class II integrons and their contribution to antibiotic resistance within *Enterobacteriaceae* family in India. *Indian J. Med. Microbiol.* 34: 303–307.

Stalder T., Barraud O., Casellas M., Dagot C. and Ploy M.-C. 2012. Integron involvement in environmental spread of antibiotic resistance. *Front. Microbiol.* 3: 119.

Tauch A., Gotker S., Puhler A., Kalinowski J. and Thierbach G. 2002. The 27.8-kb R-plasmid pTET3 from *Corynebacterium glutamicum* encodes the aminoglycoside adenyltransferase gene cassette aadA9 and the regulated tetracycline efflux system Tet 33 flanked by active copies of the widespread insertion sequence IS6100. *Plasmid* 48: 117–129.

Vaisvila R., Morgan R.D., Posfai J. and Raleigh E.A. 2001. Discovery and distribution of super-integrons among pseudomonads. *Mol. Microbiol.* 42: 587–601.

van Hoek A.H.A.M., Mevius D., Guerra B., Mullany P., Roberts A.P. and Aarts H.J.M. 2011. Acquired antibiotic resistance genes: An overview. *Front. Microbiol.* 2: 1–27.

Walsh F.M. and Amyes S.G.B. 2004. Microbiology and drug resistance mechanisms of fully resistant pathogens. *Curr. Opin. Microbiol.* 7: 439–444.

Wright G.D. 2011. Antibiotic resistome: A framework linking the clinic and the environment. In: Keen P.L. and Montforts M.H.M.M. (Eds.), *Antimicrobial Resistance in the Environment*. John Wiley & Sons, New York, pp. 15–27.

Yu W., Li Y., Liu X., Zhao X. and Li Y. 2013. Role of integrons in antimicrobial resistance: An overview. *Afric. J. Microbiol. Res.* 7: 1301–1310.

Xu Z., Li L., Alam M.J., Zhang L., Yamasaki S. and Shi L. 2008. First confirmation of integron-bearing methicillin-resistant *Staphylococcus aureus*. *Curr. Microbiol.* 57: 264–268.

Xu Z., Li L., Shirtliff M.E., Peters B.M., Li B., Peng Y., Alam M.J., Yamasaki S. and Shi L. 2011. Resistance class 1 integron in clinical methicillin-resistant *Staphylococcus aureus* strains in southern China, 2001–2006. *Clin. Microbiol. Infect.* 17: 714–718.

5 Electrophoresis
A Conceptual Understanding

Gurjeet Kaur and Vineet Awasthi

CONTENTS

5.1 INTRODUCTION

Arne Tiselius in 1931 developed a new instrument called Tiselius apparatus for moving boundary electrophoresis for separation and chemical analysis of the charged particles when subjected to electric field. The work was supported by the Rockefeller Foundation. With the evolution, the technique became more sensitive and effective and was known as zone electrophoresis. Later, the support media used was in the form of gels or filter paper, thus making electrophoresis and its related techniques an earnest tool in the bioscientific laboratories. With the advancement in the newer versions of techniques for the study of biomolecules, electrophoresis has never lost its momentum of usage in the separation of macromolecules. Electrophoresis is used as an individual as well as a supporting technique in many ways.

5.2 PRINCIPLE

Electrophoresis is based on the principle of movement of charged particles through a solution under the influence of electric field. It is the electromotive force that is used to move the molecule through the support medium. Every macromolecule possesses some net charge (q). When the electric field is applied on the molecules, the electric field strength, E, is experienced by the molecule. The molecules move toward the oppositely charged electrodes. If the molecule has net positive charge (cations), then it migrates toward cathode (cationic [negative] electrode); similarly, if the molecule has net negative charge, it migrates toward anode (anionic [positive] electrode). The migration of the molecule depends on the driving force and the resisting force experienced by the molecule. The support medium is known to suppress the thermal convection resulting due to the electric field applied, thereby retarding the movement of the molecule through the sieves formed by the support media. It is also clear that the sieves formed allow the molecule to stay at the point of separation and are ultimately stained to be visualized. The migration through the sieves of the support media allows the smaller molecules to migrate faster in comparison with the larger molecules. Thus, it is clear that the separation with the help of electrophoresis is based on size and charge possessed by the molecule. The support media commonly referred to as gels are prepared in certain percentage. The change in percentage is responsible for the pore size of the gel. The size of the molecule to be separated suggests the percentage of the gel formed for conducting electrophoresis. Higher the percentage of the gel, the smaller will be the pore size (inversely proportional). It can also be framed that bigger the molecules to be separated, smaller will be the percentage of the gel (directly proportional).

The equations necessary for better understanding of the principle of electrophoresis are mentioned as follows:

A potential gradient (E) is generated across the electrodes when the potential difference (V) is applied. Thus, the q coulomb of force carried by the molecule becomes Eq newtons. This force Eq drives the molecule to move toward its respective electrode, depending on the net charge possessed by the molecule. The support medium exerts a frictional force, thereby retarding the migration of the molecule. The retardation of the molecule is due to the shape and the hydrodynamic size of the molecule to be separated, the pore size of the separating medium, and the viscosity of the buffer (solution) used. Thus, the velocity, V, of the molecule to be separated is

$$V = \frac{Eq}{f}$$

where f is the frictional coefficient.

The other most commonly used terms are as follows:

Electrophoretic mobility (μ): As discussed earlier, the separation of the molecule is based on the size and shape of the molecule, and the friction is observed due to the support medium. These factors govern the migration or the electrophoretic mobility of the molecule. The electrophoretic mobility is expressed as follows:

$$\mu = \frac{V}{E} = \frac{q}{f}$$

The current is discontinued before the molecules reach the respective electrodes, thus ensuring the separation based on the electrophoretic mobility.

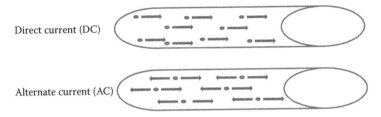

FIGURE 5.1   Visual representation of direct current and alternate current.

The potential gradient developed between the electrodes is mainly conducted by the buffer ions, and some conduction is due to the ions carried by the molecules. According to Ohm's law:

$$\frac{V}{I} = R$$

where:
   V is the voltage applied
   I is the current supplied
   R is the resistance

As a result of the process, the watts generated (W) in the support medium are given by:

$$W = I^2R$$

The equations show that the speed of the migrating molecules can be regulated by regulating the voltage applied and the current flow. The distance covered by the molecules is directly proportional to the current and time. A constant flow of the current can be maintained by using a stabilized power supply unit. It is important to note that the current applied for electrophoresis is direct current (DC) (Figure 5.1). In case of DC, the current flows in a single direction.

In alternate current, the charged molecules to be separated do not move with the applied current, as after one cycle, the poles change continuously with the changing frequency. It is experienced that on increasing the current, the sample runs faster, but consequently, the gel melts or charring at the corner of the gel takes place. Thus, to avoid such incidences, the process is carried out under cold conditions, which include performing electrophoresis in cold room or using cold running or electrode buffer.

5.3   AGAROSE GEL ELECTROPHORESIS

Agarose gel electrophoresis (AGE) is a type of gel electrophoresis used in biochemistry, molecular biology, genetics, and clinical chemistry. Agarose and agaropectin are the components of agar (Figure 5.2). Agarose is a polysaccharide isolated from

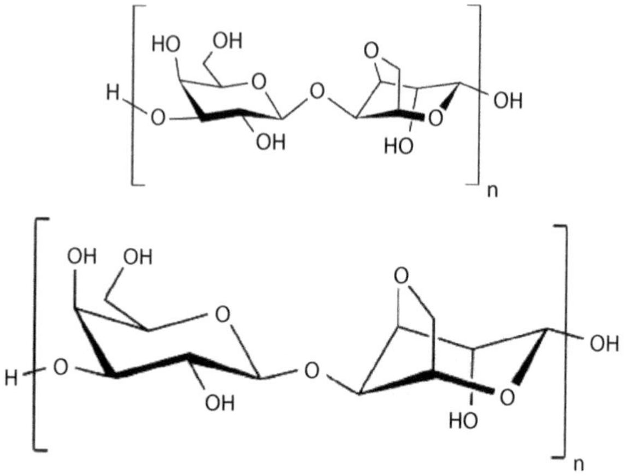

FIGURE 5.2 Structure of agarose and its components. (Courtesy of www.scienceofcooking .com/chemical_physical_properties_agar.htm.)

seaweeds. The polysaccharide is composed of agarobiose, which is the alternate units of D-galactose and 3,6-anhydrogalactose bound by (1→4) glycosidic bonds (Figure 5.2).

Agarose is a white powder, nontoxic, and neutral in nature. It is commercially available in different types, based on its purity. The purity of the agarose depends on the amount of sulfate present. Lower sulfate content confirms higher purity of agarose. The purity affects the melting temperature of agarose, strength of the gel, resolution of the separating molecule, and also the process of electroendosmosis (EEO). This process involves the electrostatic binding between the negatively charged molecules of the gel and positively charged counterions of the solvent. The positively charged ions move toward their respective electrodes when the electric field is applied. Agarose is known to transport its water of hydration, thus increasing the electroendosmotic water flow in that direction. The EEO decreases semi-logarithmically with the concentration of the gel and is analogous to the electrophoretic mobility. The process of EEO does not take place in highly pure agarose, owing to a lack of charged groups.

Agarose is commonly used as the support medium for the separation of nucleic acids and large proteins. In general, the agarose gel is used in 0.8%–1.5% concentration. The thickness of the gel made from agarose for electrophoresis ranges from 3 to 5 mm. The percentage of the gel is selected according to the size of the molecule to be separated (Table 5.1). Sieves made from the agarose gel are too large to separate small molecules such as proteins.

Agarose gel is used as a support medium in techniques such as immunoelectrophoresis and isoelectric focusing (IEF), where the separation of the proteins takes place on the basis of their native charge.

Tris–borate–ethylenediaminetetraacetic acid (EDTA) (TBE) or Tris–acetate– EDTA (TAE) could be used as the mobile media or electrode buffer. This technique helps in identification of the molecular size and weight and in separation of the charged particles.

TABLE 5.1

The Percentage of Agarose Gel Matched Base Pair for Separation

Percentage of the Agarose Gel (Wt/Vol)	Size of Nucleic Acids (in Base Pairs) to Be Separated
0.8	700–9000
1.0	500–7000
1.2	400–5000
1.5	200–3000
2.0	100–300

The advantage associated with the agarose gel is its low melting temperature of 62°C–65°C. The molecules separated could be recovered from the gel by excising the desired band and then reliquifying the gel. This results in the desired fragment recovery through AGE.

The handling of the agarose gel requires care because of its poor elasticity, due to which the gel breaks easily. This can also be explained as the gel is a combination of inter- and intramolecular hydrogen bonding between the long agarose chains. These gels run horizontally. The advantage associated with the agarose gel is that it can be reused. For reusing the gel, the gel once run is cut into small pieces and boiled with the respective buffer until completely dissolved and transparent. Once it becomes lukewarm, ethidium bromide is added and the gel is casted.

Agarose gel has direct and indirect applications. The relative molecular weight of the biomolecules can be determined when compared with the weight markers. The desired bands can be excised for further advanced experiments. The sample can be analyzed after polymerase chain reaction (PCR) and also used as a diagnostic tool in DNA fingerprinting. DNA sequencing can be performed on agarose gel by the Sanger dideoxy method. Restriction digestion mapping can also be conducted with the help of AGE.

5.4 PULSED-FIELD GEL ELECTROPHORESIS

Pulsed-field gel electrophoresis (PFGE) is a sensitive technique used for the separation of large DNA molecules (more than 50 kb), which cannot be separated by normal AGE. In this technique, the electric field of alternating polarity is given, which periodically changes the direction. The pulse time varies from 0.1 to 1000 s, which can still be variable. The movement is typically represented as *zigzag movement*. The equipment used to conduct PFGE are vertical pulsed field system and contour-clamped homogeneous field (CHEF). Owing to its property of being a highly discriminative molecular typing technique, PFGE gained impetus worldwide for the epidemiological studies. In the recent studies from past decade, the PFGE has emerged as an instrumental technique in the prevention and monitoring of diseases in different populations. It is a promising tool for epidemiological assessment, prevention, and control.

5.5 POLYACRYLAMIDE GEL ELECTROPHORESIS

Polyacrylamide gel electrophoresis (PAGE) is a competent separation technique for comparatively smaller biomolecules. The gel is made by the polymerization of acrylamide and bis-acrylamide (N,N'-methylene bis-acrylamide) (Figure 5.3).

Bis-acrylamide is used in a very small quantity, varying from 0.8% to 1.2%. Acrylamide molecules combines each other in head to tail fashion, whereas bis-acrylamide imparts the cross-linking to the gel. Owing to this, the gel becomes elastic and is thus easy to handle. The general percentage of the PAGE gel varies from 5% to 15%, keeping in mind the size of the molecule to be separated. The percentage of the gel can be varied by changing the concentration of the acrylamide and also by varying the ratio of bis-acrylamide to a minor extent. To explain it further, the gel preparation solution (commonly referred to as a stock solution) is in the ratio of 29:1 (acrylamide and bis-acrylamide). In some experiments, it can alter up to 30.2:0.8.

The polyacrylamide gel is formed by undergoing the process of polymerization. Polyacrylamide with persulfate ions of the ammonium persulfate (APS) acts as an initiator of free radicals, and N,N,N',N'-tetramethylenediamine (TEMED) acts as a catalyst for the generation of free radicals by decomposing the persulfate ions in the process of polymerization. The chemicals used for the gel preparation are hazardous, whereas after the polymerization reaction has taken place, the gel formed is not harmful. The gel casted is a result of very fast reaction taking place, as there is an involvement of free radicals, which react very fast to stabilize the lone pair of electron present in the

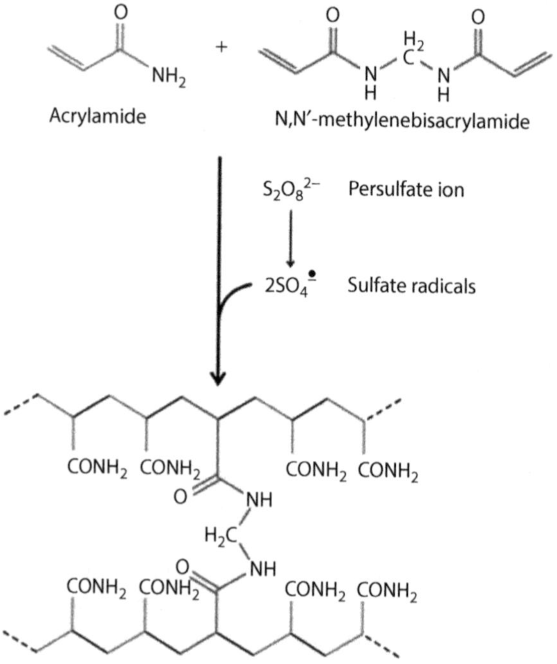

FIGURE 5.3 Structure of a polyacrylamide gel. (From http://elte.prompt.hu/sites/default/files/tananyagok/IntroductionToPracticalBiochemistry/ch07s03.html.)

outermost orbit. The process of polymerization liberates heat, and thus, air bubbles are formed during the process of gel formation. It is important to ensure that no air bubbles are trapped within the gel. Preventive measures include degassing the solutions before casting the gel, and still if any air bubble is trapped, it can be removed by carefully tapping the glass plates and allowing the bubble to move upward and out of the gel. The gel should always be handled with gloved hands to avoid impression of the hand proteins on the gel. An alternate method of photopolymerization can also be opted by using riboflavin and by subjecting the reaction to take place in the presence of light.

The PAGE gels are thinner (0.75–1.5 mm) than the agarose gels and are run vertically. These gels cannot be reused for running electrophoresis; instead, they can be restained by using the same or different staining solutions to achieve better results in contrast with the background. Most common staining solutions are Coomassie Brilliant Blue R250 (CBB-R250) and silver nitrate stain. The sensitivity of staining by CBB-R250 is 100 ng of protein and that by the silver stains is 1 ng of protein.

Adding to the advantages of PAGE, these can be stored under favorable conditions of 4°C–10°C in 30% glycerol in the storage media. The bands observed on the gel can be excised and used for further studies, using various other techniques. The PAGE gels can be used for native and chemically altered molecules. The PAGE gels are used in many different techniques of molecular biology, biochemistry, biotechnology, and genetics. The most common techniques are native PAGE, sodium dodecyl sulfate (SDS)–PAGE, Western blotting, IEF, and gel sequencing.

5.6 NATIVE PAGE

Through native PAGE, one can identify the protein in its original form (unchanged conformation). No such conditions or chemicals are used that can alter the basic nature or behavior of the proteins, and thus, the enzyme remains biologically active without undergoing any change. To the sample, low-molecular-weight dye (bromophenol blue), along with glycerol, is added to observe the run (dye front) during electrophoresis and is thus commonly termed the tracking dye. Owing to its low molecular weight, the dye observes minimum friction and runs faster than the sample containing different protein(s). This technique is used to determine the molecular weight of the protein. To identify a particular protein from a complex mixture, the gel containing protein bands is incubated with the tagged substrate, and thus, a colored product in the form of a band is observed. Another way is to lay the agarose gel over the PAGE gel containing protein bands and thus is allowed to undergo the reaction, resulting in the colored product formation. The identified gel band can be excised and used for advanced studies. Gradient gels in native PAGE are known to give better resolution for the separation of complex mixture of proteins.

5.7 SODIUM DODECYL SULFATE–POLYACRYLAMIDE GEL ELECTROPHORESIS

Sodium dodecyl sulfate–PAGE is commonly referred in short as SDS–PAGE. The proteins are separated by this technique based on their molecular weight, and the electrophoretic mobility is based on the charge-to-mass ratio. This technique is used to assess the purity of a protein during the protein purification process. The subunits of protein can also be analyzed and further confirmed by the IEF. Sodium dodecyl sulfate is an anionic detergent and has great commercial value. The other name for SDS is *lauryl sulfate*. It binds with the protein through hydrophobic interactions following stoichiometry. Sodium dodecyl sulfate, when added, imparts approximately constant charge-to-mass ratio and thus migrate at a rate proportional to its mass in the gel. In this technique, the protein is separated under the denaturing condition and thus treated with SDS to impart the overall negative charge to the protein. Beta-mercaptoethanol is added to reduce the disulfide bridges of the protein. Small-molecular-weight anionic dye is added to the sample, so that it moves faster toward the anode than the sample proteins. The dye indicates the movement of the protein within the gel. On an average, one molecule of SDS binds with two amino acid residues. The salient feature of protein *folding* is not possible, as the overall negative charge causes repulsion of the like charges. As a result, the protein becomes rod shaped in structure and anionic in nature. Hence, the protein now has changed conformation and has a rod-shape structure. This composition allows the protein to differentiate into its monomeric units. The monomeric units of the protein, if homogenous, can be further differentiated by two-dimensional (2D)-PAGE and IEF. In case of heterogeneous nature, they can be identified by SDS–PAGE. For this experiment, a discontinuous system is followed, where two gels (stacking and separating gels) are casted in a single vertical gel casting unit. Stacking gel is formed on top (not overlaid) of the separating gel in the ratio of 2:8. The stacking gel ranges from 4% to 5%, and the separating gel varies from 7% to 15%. The percentage of the separating gel is dependent on the size of the molecule to be separated (Table 5.2).

Like native gel, for better resolution and separation, the separating gel in SDS–PAGE can be prepared in the form of gradient gel. The percentage of the gel can be narrow (e.g., between 7% and 8%) or can be broad (e.g., between 8% and 11%). Several factors affect the electrophoretic mobility of the proteins, such as size of the molecule, pK value of the charged groups, concentration and pH of the buffer, temperature, friction of the gel, field strength, and the type of the support material used. The separation is based on the process called *isotachophoresis*. In this technique, both the molecule to be separated and the electrolyte move at the same speed, despite different mobility. The electrolyte moves with higher mobility as compared with the trailing ions with lower mobility. This difference is responsible for the separation of components. There is difference in the ionic strength and pH of the stacking and separating gels, thus regulating the concentration and avoiding diffusion. Thus, sharp bands are observed.

TABLE 5.2

The Percentage of Polyacrylamide Gel Matched Size-Based Protein Separation

Percentage of the Acrylamide Gel (%)	Size of the Protein to Be Separated (kDa)
8	25–200
10	15–100
12	10–70
15	6–60

It is important to stain the gel immediately after the electrophoretic run, as the protein bands tend to diffuse, resulting in blur bands rather than sharp fine bands of the separated protein. The diffusion takes place because the protein is soluble in the buffer. In the electrophoretic techniques, where the protein separated is in very low quantity, fixing the proteins with acetic acid and methanol should be done to avoid solubilization. Another reason for observed blur bands points to the fast running of the samples due to high voltage applied during electrophoresis. The CBB-R250 is commonly used to stain the bands of proteins approximately 100 ng. Alternatively, the gel can be fixed by using trichloroacetic acid (TCA) and stained by silver staining method, which stains 1 ng of protein. The process of staining is complete when the color of the stained band proteins is in the best contrast to the background of the gel. These stains bind with the protein and are visible under the white light or visible region. Colloidal stains with increased sensitivity are also available commercially.

Post staining, the relative molecular weight of the unknown protein can be determined by comparing the weight of unknown band with the known-weight markers running parallel in the well. The migration distance of the protein is measured (in cm) from the wells for each band of known and unknown proteins. The distance traveled by the dye front is also measured. The relative mobility of the protein can be calculated by dividing the protein migration distance with the distance traveled by the dye front. A graph is plotted with the relative mobility value of the known markers along the X-axis and the log of molecular mass along the Y-axis. Similarly, the relative mobility value for the unknown is calculated on the X-axis of the graph, and then, a tangent drawn on the Y-axis will give the molecular mass in log. Converting the log value will provide the molecular mass of the unknown protein (Figure 5.4).

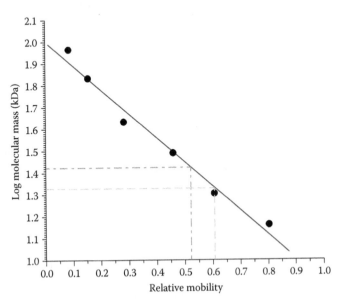

FIGURE 5.4 Molecular weight identification of unknown sample with the help of known-molecular-weight markers through graph.

Sodium dodecyl sulfate–PAGE has many applications such as peptide mapping, estimation of size and identification of molecular weight, determination of subunits of the protein, assessment of the purity and integrity of protein, and comparative analysis of number and size of the polypeptide units. It is used in Western blotting, detection of protein ubiquitination, selective labeling of cell-surface proteins using CyDye DIGE Fluor minimal dyes, and SDS–PAGE/immunoblot detection of Aβ multimers in human cortical tissues, using antigen–epitope retrieval.

5.8   GRADIENT GELS

For preparing the percentage gradient in the gel, the concentration of the support media is regulated. The gradient is prepared from larger pore size (smaller percentage) to smaller pore size (bigger percentage) of the gel. The advantage of making the gradient gel is observing better resolution of the separating molecules, based on the size. The smaller molecules move farther from the bigger ones. The gradient gel increases the range of molecular weights, and thus, distinct separation of close-molecular-weight proteins takes place.

5.9   ISOELECTRIC FOCUSING

Isoelectric focusing is a type of electrophoresis where the macromolecules are separated based on their isoelectric points (pIs). The pI is the pH at which the net charge on the protein becomes zero. In IEF, the polyacrylamide gel is mixed with the ampholytes for the separation. Current of 2500 V is applied to make pH gradient of the ampholytes across the gel. These ampholytes arrange themselves between the two electrodes, depending on their pI. The ampholytes having low pI, that is, more number of carboxylic group and a net negative charge, will move toward the anode and those with more amino group, that is, having a positive charge, will move toward the cathode. Ampholytes can be used in both broad (e.g., pH 6–10) and narrow (e.g., pH 6–7) ranges of pH. The reduced thickness of the gel is approximately 0.15 mm; this also facilitates less usage of the ampholytes, which are expensive to use. Thus, very low concentration of the sample is loaded on the gel, and depending on the pI, the protein moves on the gel and ceases at its respective pI. The advantage associated with IEF is high-resolution separation. The bands observed are very sharp, because the gradient of pH formed is very narrow. The process of staining in IEF is carried on after fixing the protein in the gel by 10% (v/v) TCA. Staining procedure of this gel is similar to the SDS–PAGE staining. The IEF is a very sensitive method of separation, as it separates the molecules at a difference of two places after decimal. Pertaining to its sensitivity, this technique is able to differentiate between the isozymes and microheterogeneity; for example, phosphorylation and dephosphorylation in a protein differ by a difference of only one or two amino acids. Relative pI of the unknown protein can also be calculated by plotting a graph. The distance traveled by each protein in the form of band is measured from a particular electrode. Distance traveled by the unknown protein band is measured, and by drawing a tangent line along the standard, the pI value for the unknown can be calculated.

The applications of IEF are very varied. In 2D electrophoresis, IEF is the first step of the two dimensions other than SDS–PAGE.

Some applications of the IEF used for various studies are mentioned as follows:

- Isoenzymes
- Variants of hemoglobin
- Proteins or peptides
- Microheterogeneity of proteins
- Genetic marker typing
- Detection of oligoclonal bands in gamma-globulin
- Probe for characterization of ionizable groups on the cell surface
- Measuring cell-surface charge densities
- pK values of ionizable groups on the cell membrane
- Following chemical modifications of the charged groups on the cell envelope

The advantages associated with the usage of IEF are reduction in the analysis time, enhanced resolution of protein bands, the possibility of subtyping the existing phenotypes, manifold increase in the sensitivity of detection, and a remarkable reduction in the counteraction of diffusion effects and thus cost per sample.

5.10   TWO-DIMENSIONAL ELECTROPHORESIS

A major challenge for the researchers undergoing investigation in the area of identification and characterization of protein can be met with the help of 2D electrophoresis. The 2D electrophoresis is considered a major tool in the area of proteomics. In the first dimension, the protein is separated based on its isoelectric point through IEF and then by SDS–PAGE. Thus, protein gets separated on the basis of both charge and molecular weight. The two dimensions undertaken in a single electrophoretic run enhance the resolution manifolds. Thousands of protein spots are observed in single 2D electrophoresis. The protein spots are stained for visualization. Various computer-based software are available to evaluate the quantitative and qualitative properties of the protein spots available on the gel. Commercially, BioNumerics 2D, Delta2D, ImageMaster 2D, Melanie, PDQuest 2-D, Samespots, REDFIN, RegStatGel, SameSpots, Gel2DE, Flicker, DeCyder 2-D, and so on, are made available by many companies for the convenience of the analysts. These software contribute in understanding the proteomic differences between the early and advanced stages of the diseased condition and achieving better resolution and differentiation between the two spots on the gel. By quantifying the protein spots, the biomarkers can be analyzed and in-depth study of the proteome can be done. Protein profiling is one of the major applications of 2D electrophoresis, which determines the presence or absence, regulation, and modification states of the proteins of interest.

5.11 CAPILLARY ELECTROPHORESIS

The assembly required to conduct capillary electrophoresis (CE) is different from the other types of electrophoresis. Based on the salient properties of CE, the other names of this technique are free solution capillary electrophoresis (FSCE), capillary zone electrophoresis (CZE), and high-performance capillary electrophoresis (HPCE). It is conducted in a long, fine, thin capillary with inner diameter ranging from 10 to 100 μm and length from 20 to 200 cm.

The sample in the capillary tube unit is injected by either of the two ways:

1. *High-voltage injection*: The positive-electrode end of the capillary tube is replaced from the reservoir container to the sample container. High voltage is then applied for a short time, until the sample rises into the capillary. The sample container is then replaced with the reservoir container, and the process of separation is continued.
2. *Pressure injection*: The positive-electrode end of the capillary tube is dipped into the air-tight, sealed sample solution container. The pressure is generated in the sample with the help of another tube inserted into the sample from the other end with the help of pressure pump. This forces the sample to rise in the capillary. The reservoir buffer is then placed back into the capillary, and the electrophoresis is conducted for separation and analysis.

The charged particles in the capillary tube move at different rates toward their respective electrodes under the influence of the electric field generated as a result of application of high voltage. It is observed that all the molecules tend to move toward the cathode, irrespective of the charge they carry. It is due to electroosmotic flow (Figure 5.5).

The electroosmotic flow is induced by the Coulomb force generated by the electric field on net mobile electric charge in a solution. An electric double layer or Debye layer of the mobile ions is formed at the interface of the solid surface and the electrolyte solution, acquiring a net fixed electrical charge. On application of the electric field, the net charge in the electric double or Debye layer is induced and it tends to move as a result of the Coulomb force generated. Thus, all the ionic molecules are dragged toward the cathode, irrespective of the charge they possess. The positively charged molecules move the fastest compared with the other charged molecules. The positively charged molecules observe both the electrophoretic mobility due to the charge they possess and electroosmotic flow effect. As they approach toward the cathode, they pass through the ultraviolet (UV) detector, which transmits

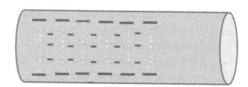

FIGURE 5.5   Schematic diagram representing electroosmotic flow in capillary electrophoresis.

the signals to the recorder, integrator, or computer. The signal is represented on the chromatograph.

The advantages associated with this technique are as follows:

- The technique is automated and hence easy to perform.
- The thin capillary solves the problems of convection current in free electrophoresis.
- The reagents for conducting the experiments are required in microliters.
- The sample for separation and analysis is required in nanoliters.
- The heat generated during the run is dissipated due to large surface area (i.e., surface-to-volume ratio), and hence, diffusion does not take place. This allows the user to subject the electrophoretic run at a higher voltage and thus reduced time.
- The sensitivity is up to femtomoles (10^{-15} moles).

Application areas include:

- Analysis of DNA, pharmaceuticals and agrochemicals
- Chiral separations
- Forensics include the analysis of illicit drugs, heroin (solid-phase extraction), hair, blood etc.
- Liquid–liquid extraction
- Diagnosis of neoplastic disorders Diagnosis of hereditary disease
- Prenatal testing
- Diagnosis of infectious disease
- Hemoglobin electrophoresis for abnormal hemoglobin detection and characterization
- Immunotyping for monoclonality
- Protein electrophoresis
- High-resolution multifraction human serum proteins
- Carbohydrate-deficient transferrin includes chronic alcohol abuse
- Molecular diagnosis includes DNA sequencing
- Analysis of DNA fragment length/restriction patterns/microsatellites
- Analysis of single-strand polymorphism.

BIBLIOGRAPHY

Anonymous. 2011. Its life science news, next biotechnology news e-bioworld. Retrieved from http://www.ebioworld.com/2011/05/polyacrylamide-gels.html.

Blaberh M. 1998. *Molecular Biology and Biotechnology*. Spring 1998. Retrieved from http://www.mikeblaber.org/oldwine/bch5425/lect20/lect20.htm.

Cheng X., Shao Z., Li C., Yu L., Raja M.A. and Liu C. 2017. Isolation, characterization and evaluation of collagen from jellyfish *Rhopilema esculentum Kishinouye* for use in hemostatic applications. *PLoS One* 12: e0169731.

Eldria K.D. and Alder D. 2004. Overview of the status and applications of capillary electrophoresis to the analysis of small molecules. *J. Chromatogr. A* 1023: 1–14.

Freifelder D. 1982. *Physical Biochemistry: Applications to Biochemistry and Molecular Biology* (2nd ed.). WH Freeman, New York.

Glynn J.R. Jr., Belongia B.M., Arnold R.G., Ogden K.L. and Baygents J.C. 1998. Capillary electrophoresis measurements of electrophoretic mobility for colloidal particles of biological interest. *Appl. Environ. Biotechnol.* 64: 2572–2577. Retrieved from https://www.labce.com/spg206950_ief_advantages_and_applications.aspx.

Julia D.W., Lance A.L. and Emanuel F.P. 2003. Proteomic applications for the early detection of cancer. *Nat. Rev. Cancer* 3: 267–275.

Kondo T. and Hirohashi S. 2009. Application of 2D-DIGE in cancer proteomic towards personalized medicine. *Methods Mol. Biol.* 577: 135–154.

Laemmli U.K. 1970. Cleavage of structural proteins during the assembly of the head of bacteriophage T4. *Nature* 227: 680–685.

Lee P.Y., Costumbrado J., Hsu C.Y. and Kim Y.H. 2012. Agarose gel electrophoresis for the separation of DNA Fragments. *J. Vis. Exp.* 62: 3923.

Mostovenko E., Hassan C., Rattke J., Deelde A.M., Van Veelen P.A. and Palmblad M. 2013. Comparison of peptide and protein fractionation methods in proteomics. *EuPA Open Proteom.* 1: 30–37.

Murch R.S. and Budowle B. 1986. Applications of isoelectric focusing in forensic serology. *J Forensic Sci.* 31: 869–880.

Parizad E.G., Parizad E.G. and Valizadeh A. 2016. The application of pulsed field gel electrophoresis in clinical studies. *J. Clin. Diagn. Res.* 10: DE01–DE04.

Razzak A. 2016. Algal Cell walls. Retrieved from http://istudy.pk/wp-content/uploads/2016/10/agarose.jpg.

Richard V.G. 2010. Pulsed field gel electrophoresis: A review of application and interpretation in the molecular epidemiology of infectious disease. *Infect. Genet. Evol.* 10: 866–875.

Righetti P.G. 1983. Isoelectric focusing: Theory, methodology and applications. In: Work T.S. and Burdon R.H. (Eds.) *Laboratory Techniques in Biochemistry and Molecular Biology*, Vol. 11. Elsevier Science, Amsterdam, the Netherlands.

Sanatan P.T., Lomate P.R., Giri A.P. and Hivrale V.K. 2013. Characterization of a chemostable serine alkaline protease from *Periplaneta americana. BMC Biochem.* 14: 32.

Ultrapure™ Agarose. Retrieved from https://www.thermofisher.com/order/catalog/product/16500500.

Walker J.M. 2002. *The Protein Protocols Handbook* (2nd ed.). Humana Press, Totowa, NJ.

Walker J.M. 1984. Gradient SDS polyacrylamide gel electrophoresis. *Methods Mol. Biol.* 1: 57–61.

Wilson K. and Walker J. 2005. *Priciples and Techniques of Biochemistry and Molecular Biology* (6th ed.). Cambridge University Press, Cambridge, UK.

6 Nutrigenomics
The Future of Human Health

Atul Bhargava and Shilpi Srivastava

CONTENTS

6.1 INTRODUCTION

Nutrition, a true integrative science, is one of the most important determinants of health and has a predominant and recognizable role in health management. Several ancient civilizations of Egypt, Persia, India, and China used food as medicine to treat and prevent diseases. The food that an individual consumes can have a positive or negative bearing on one's health. Most foods comprise components having specific biological activity. Carbohydrates, fats, and proteins are the chief components of food. These are broken down into their sugar, fatty acid, and amino acid monomeric units, respectively, through the process of hydrolysis by enzymes such as saccharidases, lipases, and proteases, respectively, which are released into the intestine from the pancreas. A nutrient is defined as a fully characterized (physical, chemical, and physiological) constituent of a diet that serves as the major energy-yielding source. It is a precursor for the synthesis of molecules needed for cell growth and development, cell differentiation, defense, repair, renewal, and/or maintenance or a required signaling molecule, cofactor, or determinant of normal molecular structure/function and/or promoter of cell and organ integrity. As environmental factors, nutrients can interact with the genetic material and can influence the development of a particular phenotype. In the molecular context, nutrients are considered as signaling molecules that transmit and translate dietary signals into changes in gene, protein, and metabolite expression via the appropriate cellular sensing mechanisms. At the genomic level, nutrients and their dietary signals serve as *signatures*, which can be precisely linked to the phenotype. Most nutrients and other bioactive food constituents are detected by cellular sensor systems and, in turn, alter the expression of the genome and the production of metabolites. Nutrients can affect the genetic array provided to us by our parents by either turning on (expression and upregulation) or turning off individual genes. It is also known that DNA metabolism, repair, and expression of genes depend on a wide range of dietary components that act as cofactors or substrates in a metabolic pathway. Numerous bioactive food constituents such as vitamins, carotenoids, some polyphenols, and terpenoids have significant beneficial effects on human health and disease prevention. These active constituents reduce the process of sustained inflammation, which is accompanied during a chronic disorder. However, a number of diseases and disorders are also related to suboptimal nutrition in terms of deficits of essential nutrients, imbalance of macronutrients, or toxic concentrations of certain food compounds. Nutritional imbalances can result in metabolic disturbances that predispose individuals to various diseases such as diabetes, cancer, osteoporosis, rheumatoid arthritis, and cardiovascular diseases (CVDs). About 30%–60% of the cancers are thought to be due to nutritional factors. Numerous studies have indicated that different cancers (breast, prostate, liver, colon, and lung cancers) are intricately associated with the dietary intakes. Several bioactive components present in the food provide protection at several stages during the development of cancer and delay its progress. Data available on variations in cancer incidence between and within populations with similar dietary habits suggest that an individual's choice of food usually reflects genetic predisposition as well as differences in gene and protein expression patterns in the individual.

The sequencing of the human genome has opened up a new field of research on the interaction between genes and food called nutrigenomics. The consumption of certain dietary components can prevent some monogenic diseases such as phenylketonuria and galactosemia. Phenylketonuria is a metabolic disorder characterized by the defective phenylalanine hydroxylase (PAH) enzyme, resulting in the accumulation of phenylalanine in the blood, which drastically increases the risk of neurological damage. Phenylalanine-restricted tyrosine-supplemented diets can reduce the risk of phenylketonuria. Likewise, galactosemia arises from a recessive trait in galactose-1-phosphate uridyltransferase (GALT), leading to the accumulation of galactose in the blood, which increases the risk of mental retardation. One way of nutritionally treating this disease is to go on a galactose-free diet. Bioactive components present in fruits and vegetables such as flavonoids (quercetin, anthocyanin, apigenin, rutin, fisetin, myricetin, kaempferol, and chalcone), phenols (curcumin, epigallocatin-3-gallate, genistein, and resveratrol), aromatic isothiocyanates, allyl sulfur compounds, indoles, protease inhibitors, and plant sterols are also known to prevent or retard carcinogenesis by mechanisms such as blocking metabolic activation through increasing detoxification.

BOX 6.1   COMMON DEFINITIONS
RELATED TO OMICS SCIENCE

Genomics: The study of the functions and interactions of all genes in the genome, including their interactions with environmental factors.

Epigenomics: The study of the complete set of epigenetic modifications on the genetic material of a cell, known as the epigenome.

Proteomics: The study of proteomes (the complete collection of proteins in a cell or tissue at a given time), which attempts to determine the role of proteins inside cells and the molecules with which they interact.

Transcriptomics: The study of the complete set of RNAs (transcriptome) encoded by the genome of a specific cell or organism.

Metabolomics: The measurement of the amounts and locations of all the metabolites in a cell, the metabolites being the small molecules transformed in the process of metabolism, that is, mostly the substrates and products of enzymes.

Metabonomics: The quantitative measurement of the time-related multiparametric metabolic response of living systems to pathophysiological stimuli or genetic modification.

Microbiomics: The study of the quality, quantity, and activity of more than 100 trillion microorganisms in the human gut.

The *omics* revolution has resulted in enormous generation of data from genomics, proteomics, transriptomics, and metabolomics studies (Box 6.1).

Quantitative techniques such as real-time polymerase chain reaction (PCR) and high-density microarray analysis have enabled us to simultaneously study the entire nutrition-relevant transcriptome. These modern techniques enable us to identify the regulatory pathways through which diet influences homeostasis (Figure 6.1).

Gene expression microarray, a technique based on nucleic acid hybridization, has gained prominence in nutrigenomics research for the investigation of the interactions between genes and the bioactive components present in the food. The primary objective of microarray expression profiling is to identify the differentially expressed genes in the condition of interest. The nutritional uptake of a person significantly affects the epigenetic modifications such as DNA methylation, histone modifications, and microRNA-based gene silencing. The Human Genome Project yielded only 20,000–24,000 genes instead of the 80,000–10,000 genes that were originally expected. Site-specific differences termed single-nucleotide polymorphisms (SNPs) occur throughout the genome, on average every 300 bases. Some genes have numerous SNPs, whereas some have none. Haplotype refers to the blocks of SNPs in individuals having SNPs in multiple genes. The pattern of SNPs is a reflection of genetic heritance, and certain haplotypes may be typical of those located within a particular region. Bulk of the genetic variation in human beings is found as polymorphisms in 90% of our genome that is nonfunctional. As a result, this variability is not reflected in the phenotype. However, variation can also be present in the coding sequences of

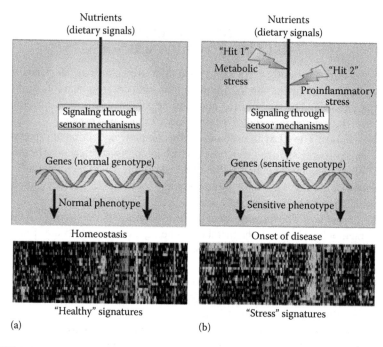

FIGURE 6.1 Comparison of healthy (a) and stress (b) signals. (Reprinted by permission from Macmillan Publishers Ltd. *Nat. Rev. Genet.*, Muller, M. and Kersten, S., 2003, copyright 2003.)

genes and in sequences regulating gene expression. This inherent genetic variation in human beings underlies the difference in susceptibility to nutrition-related traits and disorders. Even if two persons have the same diet and lifestyle, one person could develop diabetes or become obese, whereas the other would not. A new generation of techniques for high-throughput gene sequencing, also referred to as either *massive parallel* or *ultra-deep* sequencing, has arrived and offer capacity and cost structures that make it practical to study genetic variation and epigenetic marks in large populations. The next-generation sequencing technologies are the order of the day, since the concepts and procedures applied in them are aimed at increasing the sequencing throughput at a rapid pace and with less cost involved. Recent advancements in PCR and non-cloning-based technology for DNA amplification, along with improved imaging instruments, have led to the development of sequencers, which are capable of simultaneously reading millions of bases per run. Although these techniques have not been applied in nutrigenomics as of now, their future prospect is underlined by their adoption in studies of transcriptomes during physiological changes and for the comparative analysis between different disease states or conditions.

6.2   THE NUTRIGENOMICS SCIENCE

Nutrigenomics is a new branch of science that utilizes high-throughput *omics* technologies and the available genomic information to address issues related to health and nutrition. Nutrigenomics is sometimes called nutritional genomics, which is increasingly being used in a broader sense for the studies revolving around diet and health, that is, how diet affects genes and how genes affect diet. In this branch, nutrients are considered as dietary signals that influence gene and protein expressions that alter the metabolite production, and these are detected by the cellular sensor systems. In nutrigenomics, all available information about the genome and other biological molecules is vigorously utilized to unravel every detail of the interactions between the human body and the dietary intake. This emerging science of utilizing personal DNA-based information enables people to modify their nutritional behaviors, lifestyle, and nutritional supplements for serving their individual needs in the area of preventive health and aging.

There are three primary goals of nutritional genomics:

1. To establish dietary recommendations that can predict disease prevention, minimize the risk of unintended consequences, and account for the modifying effects of human genetic variation
2. To investigate the role of metabolic stress and its association with the metabolic syndrome
3. To design effective dietary regimens for the management of complex chronic disease

A comparison should also be made between nutrigenomics and nutrigenetics, both of which are intimately associated but take a fundamentally different approach to understanding the relationship between genes and diet (Figure 6.2).

Nutrigenomics aims to evaluate the effect of common dietary ingredients on the genome and attempts to relate the resulting different phenotypes to differences in the cellular and/or genetic response of the biological system. Using functional genomic tools, nutrigenomics probes a biological system following a nutritional stimulus that permits greater understanding of the role of nutritional molecules in influencing metabolic pathways and homeostatic control. However, the aim of nutrigenetics is to understand how the genetic makeup of individuals coordinates their response to diet and considers the underlying genetic polymorphisms. Nutrigenetics identifies and

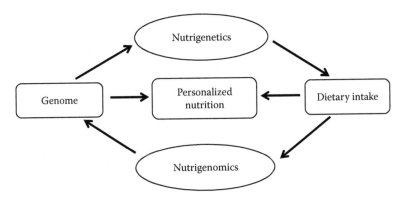

FIGURE 6.2 Relationship between nutrigenomics and nutrigenetics.

characterizes gene variants associated with differential responses to nutrients and relates this variation to the disease condition. Thus, both nutrigenomics and nutrigenetics aim to unravel genome–diet interactions albeit with different approaches and immediate goals. Nutrigenomics unravels the optimal diet from within a pool of nutritional alternatives, whereas nutrigenetics yields critically important information that will enable the clinicians to identify the optimal diet for a given individual, that is, personalized nutrition. A literature search in PubMed database using *nutrigenomics* as a keyword resulted in 1372 articles since 2005 until present, whereas 1424 articles were retrieved from PubMed database by using *nutrigenetics* as a keyword for the same time duration. Such is the potential of nutrigenomics that efforts have been made to amalgamate ancient traditional medicinal systems such as Ayurveda to this modern approach, which has given rise to a new term *Ayurnutrigenomics.* Ayurnutrigenomics is a systematic integration of nutritional practices according to Indian traditional system of medicine, Ayurveda, which integrates information from proteomics, genomics, and metabolomics. The effect of drugs and food according to the genetic constitution (*Prakriti*) of a person at the system's biology level is destined to provide a solid evidence-based scientific foundation for the advancement of personalized nutrigenomic dietetics.

6.3   TOOLS AND TECHNIQUES FOR NUTRIGENOMICS RESEARCH

Omics technologies such as genomics, transcriptomics, proteomics, and metabolomics have aided in the rapid development of nutrigenomics (Figure 6.3).

The pivotal role played by each omics technology in nutrigenomic studies is provided as follows:

1. *Genomics*: Genomics aims to decipher the structure and function of our entire DNA sequence comprising the three billion base pairs across 23 pairs of chromosomes. Genomics is concerned with the exploration of proteins and genes in the body, their activation under different conditions, and the effects of environmental factors on gene expression. Genomics employs different technologies for examining large numbers of nucleotide sequences, genes, and proteins.
2. *Transcriptomics*: Transcriptome consists of the entire complement of mRNA or transcripts generated from genes being actively transcribed or expressed. Transcriptomics is a powerful tool that is used for profiling gene expression patterns and is the most successful omics technology in nutrigenomics science, owing to its high efficiency and throughput characteristics. Genome-wide investigation of gene expression by nutrients is of prime importance in nutrigenomics research, because a wide variety of bioactive components present in the consumed food can influence the expression of genes, leading to altered biological processes. For example, microarray technology provides detailed profiles of gene expression before and after the consumption of food that facilitate simultaneous quantification of scores of mRNA. The rapid accumulation of transcriptomic microarray data in nutrigenomics has facilitated the construction of a Web-based database infrastructure built on

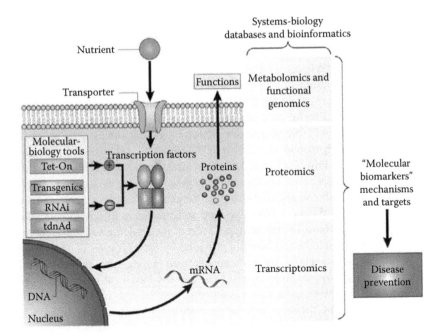

FIGURE 6.3 Use of omics technologies and molecular biology tools in nutrition science. (Reprinted by permission from Macmillan Publishers Ltd. *Nat. Rev. Genet.*, Muller, M. and Kersten, S., 2003, copyright 2003.)

an open-source database platform, which ensures the efficient organization, storage, and analysis of the enormous amount of microarray data generated from nutritranscriptomic experiments.

3. *Proteomics*: A proteome is the entire complement of proteins synthesized in a biological system at a given time and under standard set of conditions. The proteome is dynamic and more complex than the genome, since the human proteome comprises an order of about 250,000 proteins (as compared with 25,000 genes) due to alternative splicing and posttranslational modifications. Proteomics facilitates high-throughput investigation of a number of proteins, along with the discovery of novel proteins. The bottom–up approach is most widely used in proteomics, where large-scale analysis of proteomes is performed by the amalgamation of two-dimensional gel electrophoresis (2DE), followed by mass spectroscopy (MS). Shotgun proteomics is an advanced bottom–up approach that provides a rapid, detailed, and automatic identification of complex protein mixtures. The structure of proteins in the top–down approach is studied by the measurement of their intact mass, followed by direct ion dissociation in the gaseous phase, which makes it possible to distinguish between biomolecules with a high degree of sequence identity. In nutrigenomics, proteomics helps us understand how our genome is expressed as a response to diet. Integration of proteomics in nutritional science is important, since it quickly generates new data

pertaining to the complex interplay of nutrition–protein regulation, identifies novel biomarkers for nutritional status, and formulates new strategies for the prevention of diseases by regulating the diet of the individual.

4. *Metabolomics*: Metabolites are the end products of metabolic reactions, reflecting the interaction of the genome with its environment, while the metabolome consists of the entire set of metabolites synthesized in a biological system. Techniques such as MS and nuclear magnetic resonance (NMR), along with bioinformatics platforms, greatly enhance the metabolomic approach to nutrition research. Nutritional metabolomics provides greater insight into biochemical changes after dietary intervention and can impact food safety issues pertaining to genetically modified food. In nutrigenomics, metabolomics analyzes the metabolic alterations produced by the effect of nutrients or bioactive food constituents in different metabolic pathways. The most important application of metabolomics in nutrigenomics is the possible health benefits provided by the ingestion of functional compounds, especially phytochemicals such as stannols in cholesterol metabolism, flavones in heart diseases, and soy-based estrogen analogues in cancer. However, metabolism is dynamic and depends on the physiological situation and different cellular environments that makes it difficult to understand the effects separately and to directly link metabolites to genes and proteins.

6.4 IMPORTANT INITIATIVES IN NUTRIGENOMICS RESEARCH DEVELOPMENT

6.4.1 THE HUMAN VARIOME PROJECT

The human variome project (HVP) is an international effort to systematically identify genes, their mutations, and their variants associated with phenotypic variability and indications of human disease or phenotype. The main aim of HVP is to link medical, clinical, and research laboratories for developing knowledge that will be accessible to the research and medical communities worldwide to improve research strategies and clinical medical practice. The major objectives of HVP (Kaput et al. 2010) include the following:

- Capturing and archiving all gene variation associated with human disease in a central location
- Development of a standardized system of gene variation nomenclature, reference sequences, and support systems that will be easy to use by diagnostic laboratories
- Development of software to collect and exchange human variation data in gene-specific (locus-specific), country-specific, disease-specific, and general databases
- Establishing a structured and tiered mechanism that can be utilized by clinicians to determine the health outcomes associated with genetic variation

- Creation of a support system for research laboratories that would facilitate collection of genotypic and phenotypic data together, using the defined reference sequence in a free, unrestricted, and open access system
- Supporting developing countries to participate in the collection, analysis, and sharing of genetic variation information

6.4.2 THE NUTRITIONAL PHENOTYPE DATABASE

The nutritional phenotype database (dbNp) has been conceptualized as a publicly available database and knowledge repository for research related to the nutritional aspect. The objective of the dbNP is to build an extensive systems biology framework for nutrition research that would facilitate storage and retrieval of a range of relevant data, viz. preprocessed omics data, phenotype data, and study metadata. The biological information managed by the dbNP includes the data from a range of life science disciplines such as genetics, genomics, proteomics, transcriptomics, metabolomics, simple assays, functional assays, food intake, and food composition, which is of utmost importance for the needs of nutrition research. Several consortia and organizations such as Nutrigenomics Organization (NuGO), the Netherlands Metabolomics Centre (NMC), the Netherlands Bioinformatics Centre (NBIC), and the Nutrigenomics Consortium (NGC) are associated with the dbNP.

6.4.3 THE NUTRIGENOMICS ORGANIZATION

NuGO evolved from the Euorpean Union's Sixth Framework Network of Excellence and was established as an association of 23 universities and research institutes from 10 European countries (the Netherlands, Ireland, Germany, United Kingdom, Italy, Poland, Sweden, Spain, France, and Norway). It focused on jointly developing nutrigenomics and nutritional systems biology. NuGO is carrying out cutting-edge research in all the omics related to nutrition and health as well as in their ethical aspects. NuGO has now transitioned into a global association encompassing individuals and institutions around the globe. The two major objectives of NuGO (Kaput, et al. 2010) are as follows:

- Stimulating research in nutrigenomics, nutrigenetics, and systems biology relevant to nutrition and incorporation of these aspects in health research through research projects, conferences, workshops, and training programs.
- Development of bioinformatics infrastructure with reference to nutrition, by initiating, coordinating, facilitating projects in this area, along with hosting and dissemination of all data, results, and information.

6.4.4 THE HAPMAP PROJECT

Discovering the variants in DNA sequence that contributes to disease risk is of prime importance in understanding the complex causes of many common diseases in humans. Despite 99.9% similarity in DNA sequence among unrelated people, it is the remaining 0.1% that is important, because it contains the genetic variants that

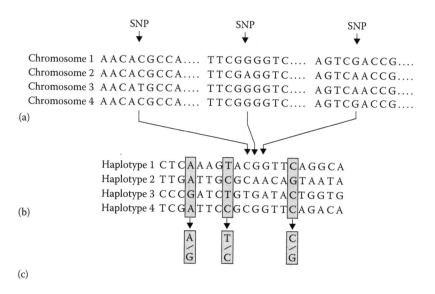

FIGURE 6.4 (a) Single-nucleotide polymorphisms, (b) haplotypes, and (c) tag single-nucleotide polymorphisms. (From The International HapMap Consortium, *Nature*, 426, 789–796, 2003.)

influence how people differ in their risk of disease or response to drugs. The SNPs are those sites in the genome where the DNA sequences of many individuals vary by a single base (Figure 6.4).

Genetic studies have indicated that about 10 million SNPs exist in human populations, and these common SNPs constitute 90% of the variation in the population. It is also interesting to note that the alleles of SNPs that are close together tend to be inherited together, which leads to associations between these alleles in the population (known as linkage disequilibrium). The specific set of alleles observed on a single chromosome, or part of a chromosome, is called a haplotype (Figure 6.4). The International HapMap Project was conceptualized to determine the common patterns of variation in the human genome at the DNA level, by characterizing sequence variants, their frequencies, and correlations in populations with diverse ancestry. The project will provide tools that will allow the indirect association approach to be applied readily to any functional candidate gene in the genome, to any region suggested by family-based linkage analysis, and ultimately to the whole genome for scans for disease risk factors. The International HapMap Project had its first meeting in October 2002, and it achieved its goal of completing the map within a short span of three years. The HapMap holds much promise as a powerful tool to enhance our understanding of the hereditary factors and the association between health and disease.

6.5 ADVANTAGES OF NUTRIGENOMICS

Nutrigenomics increases our knowledge about the mechanisms by which nutrition affects human metabolic pathways, which can be used to determine the naturally occurring chemical substances in food that could prevent the onset of diseases. Since dietary habits are known to manipulate the aging process and/or its consequences and may have potential health benefits, nutrigenomics can aid in the selection of those foodstuffs that are vital to our health. The freely available information in public domain is likely to be used by people to identify food components that are relevant to the needs of the people. Another development would be the emergence of specifically manufactured nutrigenomics food that is likely to flood the market. The possible advantages of nutrigenomics are tremendous and include the following (Ouhtit 2014):

1. A better understanding of the toxicity and safety profile of both micro- and macronutrients. The safe upper and lower limits for essential macronutrients (fats, proteins, and carbohydrates) and micronutrients (minerals and vitamins) will be better defined and understood.
2. The prevention of certain diseases that are linked to diet.
3. The consumption of otherwise-less-healthy food by individuals whose health is not likely to be affected.
4. The avoidance of unnecessary and unhelpful dietary supplements that are routinely used by certain people.
5. The prevention of diseases and an increase in life expectancy.

6.6   ISSUES, UNCERTAINTIES, AND RISKS

Although nutrigenomics offers many services directly to consumers through health-care providers, it raises several ethical, social, legal, and especially the accessibility issues with respect to how the common man will get access to nutrigenetic tests and the nutritional and lifestyle advice. The governance processes relating to nutrigenomics seem to be complicated and would require the balancing of competing interests and visions of the future. It is interesting to note that ethical and legal protections for the consumers are the major issues that would give directions to nutrigenomics in the coming decades. The research design; the collection, use, retention, and exchange of biological samples; handling of personal information; the involvement of vulnerable groups; and whether the researchers will fully report the results to consumers or their kins are likely to be the order of the day.

There are a lot of uncertainties among the consumers with respect to nutrigenomics that are not studied under risks. Data standardization and cumbersome procedures for establishing causal associations, based on sound scientific evidence, have not been up to the mark. Genes encode a variety of transcripts and proteins that have varied functions, and their interaction with the environment can lead to the formation of proteins having different or even opposing functions. Genes associated with the human body or nutrients can have diverse functions while interacting with their environment, for example, one nutrient can have a healthy effect on the heart but, at the same time, can be a risk factor with respect to cancer development. Moreover, studies on one population may not be generalized to other populations, and it is difficult to predict the validity in relation to nutrients and DNA. Another uncertainty pertains to the complexity of nutrient DNA interaction and is linked to personalized risk analysis (or diagnostic tests). Biomarkers used for diagnostic analysis make it possible to tell a person about a possible disorder related to lifestyle-related conditions in an early but reversible stage. However, such valid biomarkers are not yet readily available. Apart from this, many health vulnerabilities do not develop into full blown diseases due to normal reactions of the body, and their identification with *prediseases* is premature. This is a fundamental uncertainty that is applicable to diagnostic tests indicating a *predisease*. Uncertainties also exist with respect to the preventive measures and dietary advice based on DNA testing. It remains to be seen to what extent may the risks be reduced by eating a certain diet or by adopting a healthy lifestyle, since the preventive actions taken to delay the health risks based on predictive testing are, in many cases, not fully successful. Finally, uncertainties also exist with respect to the amalgamation of nutrigenomics data with functional food development and their acceptance by the consumers. There is considerable skepticism whether nutrigenomics will be a commercial tool or whether it be used to develop health-improvement, disease-prevention, and performance-enhancing products.

6.7 OPPORTUNITIES AND CHALLENGES

The consumers, who are major stakeholders, need nutrigenomics for fine-tuning of diets, especially in a scenario where the person does not know what is healthy to eat. With respect to basic nutrigenomics research and its commercial and clinical uses, five potential areas have been identified:

1. Health claim benefits arising from nutrigenomics research
2. Management of enormous information generated in nutrigenomics
3. Different delivery methods that would be associated with nutrigenomics services
4. Products related to nutrigenomics
5. Accessibility to nutrigenomics by all types of societies and/or classes

6.8 FUTURE OF NUTRIGENOMICS

Nutrigenomics is still in its infancy, and there are many uncertainties about its further development, from both technological point of view and acceptance by the society. A thorough understanding of the human genome is likely to further accelerate nutrigenomics science and aid in the development of nutritional modifications, including personalized nutrition, that would be crucial for our well-being. This would, in turn, lead to a strong impetus for future drug discovery. Nutrigenomics is promising for the development of functional foods and nutraceuticals. In the coming decades, nutrigenomics is likely to accelerate the development of food products, especially nutraceuticals, and provide dietary advice catering to the nutritional needs of specific groups of consumers, societies, or even individuals. The growing awareness of the prevalence of genomic profiles and studies of SNPs will enable grouping of consumers with specific genetic profiles that may predispose them to particular disease and mass customization rather than individualization of nutrigenomic foods. For nutrigenomic foods to emerge as a new category of foods, intense business planning will be required in the production, marketing, and distribution of these food products. Since nutrigenomic foods will essentially be high-value products, stakeholders in the production units must undertake extensive market research before initiating production.

BIBLIOGRAPHY

Affolter M., Raymond F. and Kussmann M. 2009. Omics in nutrition and health research. In: Mine Y., Miyashita K. and Shahidi F. (Eds.) *Nutrigenomics and Proteomics in Health and Disease: Food Factors and Gene Interactions.* Wiley Blackwell Edition, Ames, IA, pp. 11–29.

Afman L. and Muller M. 2006. Nutrigenomics: From molecular nutrition to prevention of disease. *J. Am. Diet. Assoc.* 106: 569–576.

Arab L. 2004. Individualized nutritional recommendations: Do we have the measurements needed to assess risk and make dietary recommendations? *Proc. Nutr. Soc.* 63: 167–172.

Astley S.B. 2007. An introduction to nutrigenomics developments and trends. *Genes Nutr.* 2: 11–13.

Banerjee S., Debnath P. and Debnath P.K. 2015. Ayurnutrigenomics: Ayurveda-inspired personalized nutrition from inception to evidence. *J. Tradit. Complement. Med.* 5: 228–233.

Barnes S., Prasain J. and Kim H. 2013. In nutrition, can we 'see' what is good for us? *Adv. Nutr.* 4: 327S–334S.

Berdanier C.D. and Hargrove J.L. 1993. *Nutrition and Gene Expression.* CRC Press, Boca Raton, FL.

Bergmann M.M., Bodzioch M., Bonet M.L., Defoort C., Lietz G. and Mathers J.C. 2006. Bioethics in human nutrigenomics research: European Nutrigenomics Organisation Workshop Report. *Br. J. Nutr.* 95: 1024–1027.

Bergmann M.M., Gorman U. and Mathers J.C. 2008. Bioethical considerations for human nutrigenomics. *Annu. Rev. Nutr.* 28: 447–467.

Bhargava A. and Srivastava S. 2012. Metabolomics—the technology of the future. In: Bhargava A. and Srivastava S. (Eds.) *Biotechnology: New Ideas, New Developments.* Nova Science Publishers, Hauppauge, NY, pp. 45–68.

Brennan L. 2013. Metabolomics in nutrition research: Current status and perspectives. *Biochem. Soc. Trans.* 41: 670–673.

Brennan R.O. 1977. *Nutrigenetics: New Concepts for Relieving Hypoglycemia.* Signet, New York.

Bull C. and Fenech M. 2008. Genome health nutrigenomics and nutrigenetics: Nutritional requirements for chromosomal stability and telomere maintenance at the individual level. *Proc. Nutr. Soc.* 7: 146–156.

Burke W., Khoury M.J., Stewart A. and Zimmern R.L. 2006. The path from genome-based research to population health: Development of an international public health genomics network. *Genet. Med.* 8: 451–458.

Castle D. and Ries N.M. 2007. Ethical, legal and social issues in nutrigenomics: The challenges of regulating service delivery and building health professional capacity. *Mutat. Res.* 622: 138–143.

Castro C.E. and Towle H.C. 1986. Nutrient-genome interaction. *Fed. Proc.* 45: 2382.

Chen C. and Kong A.N. 2005. Dietary cancer chemo preventive compounds: From signaling and gene expression to pharmacological effects. *Trends Pharmacol. Sci.* 26: 318–326.

Daniel H. 2002. Genomics and proteomics: Importance for the future of nutrition research. *Br. J. Nutr.* 87: 305–311.

Daniel H. and Wenzel U. 2006. Nutritional genomics: Concepts, tools and expectations. In: Brigelius-Flohé R. and Joost H.-G. (Eds.) *Nutritional Genomics.* Wiley, Hoboken, NJ.

Davis C.D. and Milner J. 2004. Frontiers in nutrigenomics, proteomics, metabolomics and cancer prevention. *Mutat. Res.* 551: 51–64.

Davis C.D. and Milner J.A. 2009. Gastrointestinal microflora, food components and colon cancer prevention. *J. Nutr. Biochem.* 20: 743–752.

DellaPenna D. 1999. Nutritional genomics: Manipulating plant micronutrients to improve human health. *Science* 285: 375–379.

Desiere F. 2004. Towards a systems biology understanding of human health: Interplay between genotype, environment and nutrition. *Biotechnol. Annu. Rev.* 10: 51–84.

Elliott R. and Ong T.J. 2002. Nutritional genomics. *BMJ* 324: 1438–1442.

El-Sohemy A. 2007. Nutrigenetics. *Forum Nutr.* 60: 25–30.

Evelo C.T., van Bochove K. and Saito J.-T. 2011. Answering biological questions: Querying a systems biology database for nutrigenomics. *Genes Nutr.* 6: 81–87.

Fenech M. 2008. Genome health nutrigenomics and nutrigenetics-diagnosis and nutritional treatment of genome damage on an individual basis. *Food Chem. Toxicol.* 46: 1365–1370.

Ferguson L.R. 2006. Nutrigenomics: Integrating genomics approaches into nutrition research. *Mol. Diag. Ther.* 10: 101–108.

Ghosh D. 2009. Future perspectives of nutrigenomics foods: Benefits vs risks. *Ind. J. Biochem. Biophys.* 46: 31–36.

Gibney M.J. and Walsh M.C. 2013. The future direction of personalised nutrition: My diet, my phenotype, my genes. *Proc. Nutr. Soc.* 72: 219–225.

Gillies P.J. 2003. Nutrigenomics: The rubicon of molecular nutrition. *J. Am. Diet. Assoc.* 103: S50–S55.

International HapMap Consortium. 2003. The international HapMap project. *Nature* 426: 789–796.

Jones D.P., Park Y. and Ziegler T.R. 2012. Nutritional metabolomics: Progress in addressing complexity in diet and health. *Annu. Rev. Nutr.* 32: 183–202.

Joost H.-G., Gibney M.J., Cashman K.D., Gorman U., Hesketh J.E., Mueller M., van Ommen B., Williams C.M. and Mathers J.C. 2007. Personalized nutrition: Status and perspectives. *Br. J. Nutr.* 98: 26–31.

Junien C. and Gallou C. 2004. Cancer nutrigenomics. Nutrigenetics and nutrigenomics. *World Rev. Nutr. Diet* 93: 210–269.

Kaptur J. and Raymond L.R. 2004. Nutritional genomics: The next frontier in the post genomic era. *Physiol Genom.* 16: 167–177.

Kaput J. 2007. Nutrigenomics—2006 update. *Clin. Chem. Lab Med.* 45: 279–287.

Kaput J. 2008. Nutrigenomics research for personalized nutrition and medicine. *Curr. Opin. Biotechnol.* 19: 110–120.

Kaput J. and Rodriguez R.L. 2004. Nutritional genomics: The next frontier in the post genomic era. *Physiol. Genom.* 16: 166–177.

Kaput J., Astley S., Renkema M., Ordovas J. and van Ommen B. 2006. Harnessing nutrigenomics: Development of web-based communication, databases, resources, and tools. *Genes Nutr.* 1: 5–11.

Kaput J., Evelo C.T., Perozzi G., van Ommen B. and Cotton R. 2010. Connecting the human variome project to nutrigenomics. *Genes Nutr.* 5: 275–283.

Kato H., Saito K. and Kimura T. 2005. A perspective on DNA microarray technology in food and nutritional science. *Curr. Opin. Clin. Nutr. Metab. Care* 8: 516–522.

Kim Y.S. and Milner J.A. 2011. Bioactive food components and cancer-specific metabonomic profiles. *J. Biomed. Biotechnol.* 2011: 721213.

Korthals M. and Komduur R. 2010. Uncertainties of nutrigenomics and their ethical meaning. *J. Agric. Environ. Ethics* 23: 435–454.

Link A., Balaguer F. and Goel A. 2010. Cancer chemoprevention by dietary polyphenols: Promising role for epigenetics. *Biochem. Pharmacol.* 80: 1771–1792.

Llorach R., Garcia-Aloy M., Tulipani S., Vazquez-Fresno R. and Andres-Lacueva C. 2012. Nutrimetabolomic strategies to develop new biomarkers of intake and health effects. *J. Agric. Food Chem.* 60: 8797–8808.

Lyons C.L., Kennedy E.B. and Roche H.M. 2016. Metabolic inflammation-differential modulation by dietary constituents. *Nutrients* 8: 247.

Mariman E.C. 2006. Nutrigenomics and nutrigenetics: The 'omics' revolution in nutritional science. *Biotechnol. Appl. Biochem.* 44: 119–128.

Micó V., Díez-Ricote L. and Daimiel L. 2016. Nutrigenetics and nutrimiromics of the circadian system: The time for human health. *Int. J. Mol. Sci.* 17: 299.

Milner J.A. 2006. Diet and cancer: Facts and controversies. *Nutr. Cancer.* 56: 216–224.

Muller M. and Kersten S. 2003. Nutrigenomics: Goals and strategies. *Nat. Rev. Genet.* 4: 315–322.

Mutch D.M., Wahli W. and Williamson G. 2005. Nutrigenomics and nutrigenetics: The emerging faces of nutrition. *FASEB J.* 19: 1602–1616.

Neeh V.S. and Kinth P. 2013. Nutrigenomics research: A review. *J. Food Sci. Technol.* 50: 415–428.

Nicastro H.L., Trujillo E.B. and Milner J.A. 2012. Nutrigenomics and cancer prevention. *Curr. Nutr. Rep.* 1: 37–43.

Nuno N.B. and Heuberger R. 2014. Nutrigenetic associations with cardiovascular disease. *Rev. Cardiovasc. Med.* 15: 217–225.

Ordovas J.M. and Corella D. 2004. Nutritional genomics. *Annu. Rev. Genomics Hum. Genet.* 5: 71–118.

Ordovas J.M. and Mooser V. 2004. Nutrigenomics and nutrigenetics. *Curr. Opin. Lipidol.* 15: 101–108.

Ouhtit A. 2014. Nutrigenomics: From promise to practice. *Sultan Qaboos Univ. Med. J.* 14: 1–3.

Palou A. 2007. From nutrigenomics to personalised nutrition. *Gen. Nutr.* 2: 5–7.

Patrinos G.P. and Prainsack B. 2014. Working towards personalization of medicine: Genomics in 2014. *Pers. Med.* 11: 611–613.

Pavlidis C., Patrinos G.P. and Katsila T. 2015. Nutrigenomics: A controversy. *Appl. Transl. Genom.* 4: 50–53.

Peregrin T. 2001. The new frontier of nutrition science: Nutrigenomics. *J. Am. Diet Assoc.* 101: 1306.

Qi L. and Cho Y.A. 2008. Gene environment interaction and obesity. *Nutr. Rev.* 66: 684–694.

Ring H.Z., Kwok P.Y. and Cotton R.G. 2006. Human Variome Project: An international collaboration to catalogue human genetic variation. *Pharmacogenomics* 7: 969–972.

Riscuta G. 2016. Nutrigenomics at the interface of aging, lifespan, and cancer prevention. *J. Nutr.* 146: 1931–1939.

Riscuta G. and Dumitrescu R.G. 2012. Nutrigenomics: Implications for breast and colon cancer prevention. *Methods Mol. Biol.* 863: 343–358.

Roberts M.A., Mutch D.M. and German J.B. 2001. Genomics: Food and nutrition. *Curr. Opin. Biotechnol.* 12: 516–522.

Ruth L. 2007. Nutrigenomics: Impacts on markets, diets, and health. Part 1: Biotechnology and diagnostics. *Genes Nutr.* 2: 21.

Ruth L. and Wrick K.L. 2005. Nutrigenomics: Impact on diets, markets and health. *Adv. Life Sci. Rep.* 48: 1–21.

Sales N.M.R., Pelegrini P.B. and Goerch M.C. 2014. Nutrigenomics: Definitions and advances of this new science. *J. Nutr. Metab.* 2014: 202759.

Schmelz E.M., Wang M.D. and Merrill A.H. Jr. 2006. Genomics, proteomics, metabolomics and systems biology approaches to nutrition. In: Bowman B.A. and Russel R.M (Eds.), *Present Knowledge of Nutrition* (9th ed.). International Life Science Institute, Washington DC, pp. 3–19.

Stover P.J. 2004. Nutritional genomics. *Physiol. Genom.* 16: 161–165.

Subbiah M.T. 2007. Nutrigenetics and nutraceuticals: The next wave riding on personalized medicine. *Transl. Res.* 149: 55–61.

Surh Y.J. 2003. Cancer chemoprevention with dietary phytochemicals. *Nat. Rev. Cancer* 3: 768–780.

Trujillo E., Davis C. and Milner J. 2006. Nutrigenomics, proteomics, metabolomics, and the practice of dietetics. *J. Am. Diet Assoc.* 106: 403–413.

van Ommen B. and Stierum R. 2002. Nutrigenomics: Exploiting systems biology in the nutrition and health arena. *Curr. Opin. Biotechnol.* 13: 517–521.

van Ommen B., Bouwman J., Dragsted L., Drevon C.A., Elliott R., de Groot P., Kaput J. et al. 2010. Challenges of molecular nutrition research 6: The nutritional phenotype database to store, share and evaluate nutritional systems biology studies. *Genes Nutr.* 5: 189–203.

Wishart D.S. 2008. Metabolomics: Applications to food science and nutrition research. *Trends Food Sci. Tech.* 19: 482–493.

Wu A.H., Yu M.C., Tseng C.C. and Pike M.C. 2008. Epidemiology of soy exposures and breast cancer risk. *Br. Cancer* 98: 9–14.

Zeisel S.H. 2010. A grand challenge for nutrigenomics. *Front. Genet.* 1: 1–3.

Zeisel S.H., Allen L.H., Coburn S.P., Erdman J.W., Failla M.L., Freake H.C., King J.C. and Storch J. 2001. Nutrition: A reservoir for integrative science. *J. Nutr.* 131: 1319–1321.

Zhang W., Ratain M.J. and Dolan M.E. 2008. The HapMap Resource is providing new insights into ourselves and its application to pharmacogenomics. *Bioinform. Biol. Insights* 2: 15–23.

7 Advances in Computer-Aided Drug Designing

Ajay Kumar Singh

CONTENTS

7.1 INTRODUCTION

Designing of a new drug and bringing it from laboratory to market is a costly, time-consuming, and manual process. The process is exhaustive in terms of formulation, prediction of targets, and development of new leads for multiple targets. It was a costly task that averaged USD 800–USD 1.8 million and took 10–15 years for the development of a new drug. Development of a new drug increases the use of new approaches and databases for ligands and targets selection process, which is leading to generation of high-throughput data screening. Computer-aided drug designing (CADD) is amalgamating new branches of science and technology, viz. biochemistry, biomedical science, chemoinformatics, and nanotechnology. At nano level to understand the approaches for drug-delivery systems. Target selection, ligand identification and screening, interaction of molecules in three-dimensional (3D) space, binding mode identification and selection, and energy calculation are now easier after using computers in drug designing and development process. In-silico drug design or CADD also plays an important role in hit identification, in hit-to-lead selection, and in finding out the optimum solutions of absorption, distribution, metabolism, excretion, and toxicity profile. This analysis helps avoid safety issues for new drug development. Computer-aided drug designing has oriented in post-genomic era for the considerable range of application for the discovery and development of pipeline of drugs from target identification to lead identification and from lead optimization to preclinical or clinical trials.

7.2   ANCIENT APPROACHES TO DRUG DEVELOPMENT

In ancient times, the traditional societies throughout the world were dependent on the traditional medicine systems as primary health resources. The uses of plants as therapeutic medicine for the treatment of most human and livestock ailments have a long history probably since 4000–5000 B.C. Plants were used by the Chinese people initially as natural herbal medicines. However, plants as natural medicinal resources have been extensively reported in the ancient Indian text the Rig Veda, which is said to be written between 1600 B.C. and 3500 B.C. The medicinal plant systems have been in practice in many countries throughout the world. The new drug molecules developed are now rich resource of ethnobotany for drug development process.

People trusted nature for remediation of various illnesses and for finding suitable cures by using plant-based products. The natural plant world shows immense potential for new drug discovery. Drug development from natural resources has been the profound field of interest to scientists who aim to find the precursors molecules used since ancient times. These natural sources were not only successful in the ancient period, but they are also being used for commercial development of new drugs (Figure 7.1).

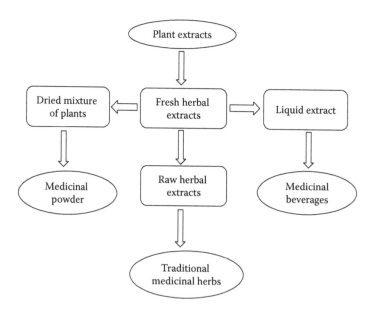

FIGURE 7.1   Ancient herbal medicinal uses.

In modern times, the development of a plant-based drug is based on hit-and-trial methods. The drawback of taking natural medicine for various ailments in crude form by the patients is the intake of many molecules apart from the molecule of interest. There is also a lack of awareness with respect to their side effects and synergistic effects. Technology development in isolation of specific compounds extraction methods has given a new direction to drug leads in the nineteenth century. Quinine is one such example. It was isolated in 1823 but has gone rapid strides since then.

7.3 DRUG DEVELOPMENT FROM NATURAL SOURCES

Ancient history of medicinal plant clearly indicated the use of clay tablets in Mesopotamia (2600 B.C.) for the use of oils from *Cupressus sempervirens* (cypress) and *Commiphora* species (myrrh). These are still in use for the treatment of colds, coughs, and inflammation. One of the famous Egyptian pharmaceutical work, the Ebers Papyrus (1550 B.C.), stated the use of 700 plant-based drugs, viz. gargles, pills, infusions, and ointments. The use of natural products is also well documented in Chinese "Pen T'Sao" (2500 B.C.; contains 365 drugs), "Materia Medica" (1100 B.C.; contains 52 prescriptions), "Shennong Herbal" (100 B.C.; contains 365 drugs), and "Tang Herbal" (695 A.D.; contains 850 drugs). About 63 plant species from the Minoan, Mycenaean, and Egyptian Assyrian pharmacotherapy have been referred to in Homer's epics, viz. the Iliad and the Odyssey, created circa 800 B.C. Pedanius Dioscorides, the father of pharmacognosy, studied medicinal plants and wrote the work "De Materia Medica," which described 657 drugs of plant origin, including the making of the medicinal preparations, and their therapeutic effect.

Arabs were the first to introduce the concept of pharmacies (in the eighth century) with the help of Avicenna. The work of Avicenna, one of the famous Persian pharmacists and physicians, *The Canon of Medicine*, a medical encyclopedia, was widely studied as a medical text at many universities during the medieval period and was in use as late as 1650. Development of current drugs was based on the traditional medicinal use of medicine in clinical, pharmacological, and chemical studies.

7.4 SYNTHESIS-BASED DRUG DEVELOPMENT

There are various drugs that have been synthesized from anti-inflammatory agents such as acetylsalicylic acid (aspirin) (Figure 7.2a). Some of the drugs have been derived from the natural products such as salicin (Figure 7.2b), which is isolated from the bark of the willow tree *Salix alba* L.

An alkaloid source named morphine (Figure 7.2c) was first reported in 1803 and was isolated from *Papaver somniferum* L., which is commonly known as opium poppy. Although *Digitalis purpurea* L. (foxglove) was used as a medicinal plant in Europe by the tenth century, its active constituent digitoxin (Figure 7.2d) was isolated only till 1700s. Active compound digitoxin was reported to be a glycoside that enhanced cardiac conduction, which in turn improved the cardiac contractibility. Digitoxin and its derivatives have been used in many ancient

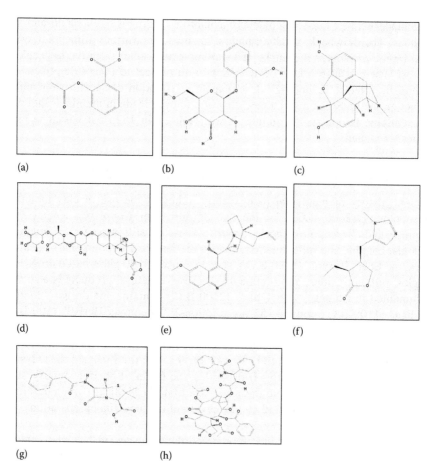

FIGURE 7.2 Synthesis-based drug molecules: (a) acetylsalicylic acid, (b) salicin, (c) morphine, (d) digitoxin, (e) quinine, (f) pilocarpine, (g) penicillin, and (h) paclitaxel.

civilizations in the management of congestive heart failure. They have long-term damaging effects on heart and are primarily used in the treatment of heart deficiency. The anti-malarial drug quinine (Figure 7.2e), isolated from bark of *Cinchona succirubra*, was also approved by the U.S. Food and Drug Administration in 2004. The medicinal use of Klotzsch is for malaria, fever, indigestion, throat and mouth diseases, and cancers. Another important drug is pilocarpine (Figure 7.2f), obtained from *Pilocarpus jaborandi* (Rutaceae), which is an L-histidine-derived alkaloid. This has been best used for the last 100 years for acute angle-closure glaucoma.

These are a few examples of synthetic medicinal compounds used as medicines in ancient times. Penicillin (Figure 7.2g), the famous natural product, was isolated from *Penicillium notatum* by Alexander Fleming in 1929. The Nobel Prize in Physiology or Medicine 1945 was awarded jointly to Alexander Fleming, Ernst Boris Chain, and Sir Howard Walter Florey for the discovery of penicillin and its curative effect in various infectious diseases. Plant-based anticancer drug paclitaxel (Figure 7.2h) was isolated from bark of *Taxus brevifolia* for breast cancer. The U.S. Department of Agriculture first collected the drug in 1962 by the plant-screening program at the National Cancer Institute.

7.5   MODERN DRUG DEVELOPMENT

The foundation of CADD was laid more than a century back, when mathematics was applied for explaining and calculating chemical property. "Next Industrial Revolution: Designing Drugs by Computer at Merck" was the pioneer article published by *Fortune* magazine on October 5, 1981. It was the first credit for the new potential drug designing process through computer assisted designing. The Computer-Aided Drug Design Center was established to initiate collaborative research between biologists, biophysicists, structural biologists, and computational scientists. This new branch of science relied on computers to screen the huge number of necessary compounds for selection of the lead compound. Since CADD uses a much more targeted search than traditional high-throughput screening (HTS) and combinatorial chemistry, it is capable of increasing the hit rate of novel drug compounds. Thus, CADD not only aims to explain the molecular basis of therapeutic activity but is also useful in predicting the possible derivatives that would improve the activity. The approach for finding new drug leads now requires less complicated validation approaches. Computer-aided drug designing significantly decreases the number of compounds necessary to screen, but at the same time, it retains the same level of lead compound discovery. The CADD's historical prospective clearly indicates its involvement in drug discovery.

- 1900: The receptor and lock-and-key concepts.
- 1970s: Quantitative structure–activity relationships (QSAR). Limitations: Two-dimensional, retrospective analysis.
- 1980s: Beginning of CADD molecular biology, X-ray crystallography, multidimensional nuclear magnetic resonance (NMR) and molecular modeling, and computer graphics.
- 1990s: Human genome bioinformatics, combinatorial chemistry, and HTS.

The current CADD approach requires various approaches to find out the lead compounds. The various steps for a lead compound can be predicted by a working model shown in Figure 7.3.

Computer-aided drug designing is unique in the sense that it makes use of computer-based methods for simulating drug–receptor interactions. The methods employed in CADD greatly rely on modern bioinformatics tools, information technology, information management, software applications, databases, and computational resources, all of which provide the infrastructure for bioinformatics.

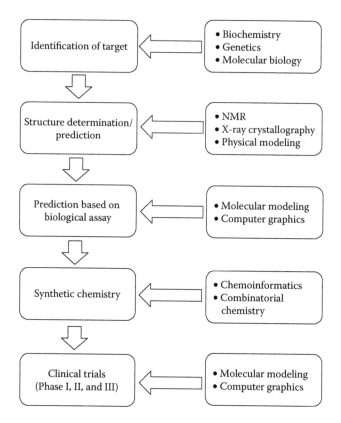

FIGURE 7.3   Graphical representation of computer-aided drug design.

7.6   STRATEGIES IN COMPUTER-AIDED DRUG DESIGNING

1. *Target identification*: Target identification is a field integrating different levels of information in drug–protein and protein–disease networks and involves an intricate network between databases and correlations across genomics, proteomics, transcriptomics, metabolomics, microbiome, and pharmacogenomics. There are two approaches that can be applied for target identification: the first one is at the start (target-based or reverse chemical genetics) and the second one is at the end (phenotype-based or forward chemical genetics) of biological screening. The selection of target can be done in three ways: biochemical, genetic interaction, and computational inferences. Biochemical affinity purification hits the target of small molecules directly through biochemical approaches. However, lead optimization is good on availability of target structure information with the help of enzymes assays and other biochemical tests. Various DNA and RNA analysis-based methods allow target identification via genetic or genomic methods for the functional analysis of protein targets in controlled system. Predictions of targets of identified inhibitors by using structure-based and profiling (ligand-based) methods have also been reported on numerous computational inference methods.

2. *Structure prediction*: During the initial stages of drug development process, researchers faced challenges due to little or no information of structure–activity relationship (SAR) information. Now, the development of HTR techniques has made it easier for the chemists to develop assay and screening of lead compounds. Selection of lead compounds is based on a set of compounds that have diversity in their physiochemical properties and are easier to select rather than random selection. The aim of these analyses is to select and test less compounds while gaining as much information as possible about the data set. The selection of compounds to be tested is of prime importance, since it has a strong impact on the number of compounds screened for research efficiency and also cost implications. Investigations for refinements and lead findings were recently investigated for the rational drug design to structural diversity of databases. Three-dimensional databases for the diversity enhancement were compared by the hierarchical clustering and maximum dissimilarity methods. The rational selection method and random approaches were compared for the performance of two-dimensional fingerprints as a validated molecular descriptor. This comparison leads to the selection of pharmacophores. Pharmacophore model is a spatial arrangement of atoms or functional groups that explains the binding of structurally diverse ligands to a common receptor site.

3. *Biological assay-based prediction for targets*: A bioassay or a biological assay is defined as a testing procedure that is used to estimate the concentration of a pharmaceutically active substance in a formulated product or bulk material. Of the different physical or chemical methods available, a bioassay yields exhaustive information on the biological activity of a substance. The CADD-based drug target identification can be dividing into drug-based similarity inference (DBSI), target-based similarity inference (TBSI), and network-based inference (NBI). The NBI method is considered the best method among them. These *in vitro* assays have proved that the five old drugs, namely *montelukast, diclofenac, simvastatin, ketoconazole, and itraconazole* show polypharmacological features on estrogen receptors or dipeptidyl peptidase-IV. Target prediction may also be performed by receptor-based methods such as reverse docking, which have also been applied in drug–target (DT) binding affinity prediction. The limitation of the target prediction is the availability of 3D structure.

7.7 COMPUTER-AIDED DRUG DESIGNING-BASED DRUG DEVELOPMENT STRATEGIES

Computer-aided drug designing strategies have been classified into two categories:

1. The molecules for which the 3D structure information and their functional role are known. This is a well-established method used by the pharmaceutical industry nowadays for the development of small drug molecules with desired properties.

2. The molecules for which the 3D structure information and their functional role may be known or unknown. Unknown drug targets may be predicted by drug designing tools.

The stages of both strategies may be understood by Figures 7.4 and 7.5. Both the strategies may extend for further analysis of many other features shown in Figure 7.5.

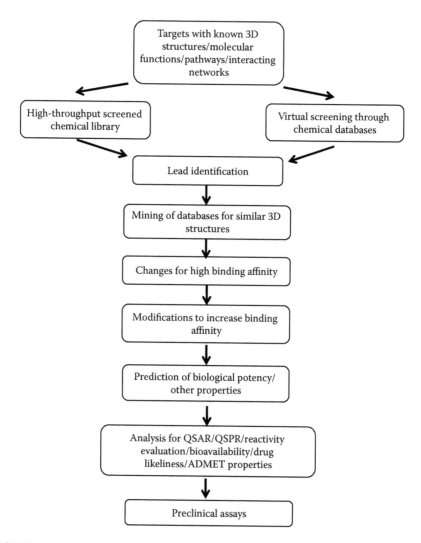

FIGURE 7.4 Stepwise process for drug designing with known target 3D structure. (Reprinted from *Eur. J. Pharmacol.*, 625, Mandal S. et al., Rational drug design, 90–100, Copyright 2009, with permission from Elsevier.)

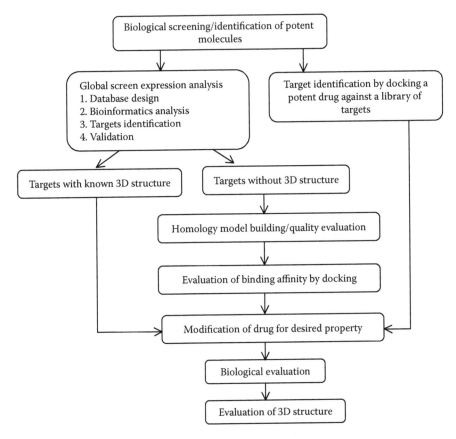

FIGURE 7.5 Stepwise process for drug designing with unknown targets. (Reprinted from *Eur. J. Pharmacol.*, 625, Mandal S. et al., Rational drug design, 90–100, Copyright 2009, with permission from Elsevier.)

The successful identification of the target is followed by exhaustive study of numerous aspects like binding scores (affinity/specificity); absorption, distribution, metabolism and excretion (ADME); biodegradation; Quantitative structure–activity relationship (QSAR) and quantitative structure property relationship (QSPR). All these features are required for the development of molecules, which is accomplished through computational tools. The role of these molecules is to find the gene expression patterns via gene expression profiling after initial evaluation and identification of lead molecules. The analysis of these parameters helps in the improvement of drug for the various aspects such as desired attributes for disease-free survival, eradication of disease, toxic side effect minimization or elimination, improvement of immune responses, overcoming of drug resistance, and improvement in distribution (bioavailability). This drug designing approach integrates to the rational drug designing approach for drug development (Figure 7.6).

Examination of
1. QSAR
2. QSPR
3. Potency
4. Docking and performance of
   multilinear regression analysis

Reactivity evaluation (examination of
biodegradation profile):
1. Electrophilic
2. Nucleophilic
3. Radical attack

Improvement of bioavailability (modified Lipinski's rules for ADME) and
examination of various drug-like properties

Evaluation of
1. *In vivo* experiments
2. Gene expression profiling
   (a) Bioinformatics analysis
   (b) Identification of genes
       responsible for:
       • Toxicity
       • Drug resistance
       • Metabolism
       • Immune suppression

Preclinical evaluation

FIGURE 7.6 Examination of additional properties toward the improvement of drug-like properties. (Reprinted from *Eur. J. Pharmacol.*, 625, Mandal S. et al., Rational drug design, 90–100, Copyright 2009, with permission from Elsevier.)

7.8   CURRENT SOFTWARE, TOOLS, AND PROGRAMS FOR IN SILICO DRUG DESIGNING

Modern drug development in the pharmaceutical industry started in the early twentieth century. Paul Ehrlich, one of the Nobel Laureates, considered as the father of modern chemotherapy, is famous for the development of new synthesis of drugs in laboratories. Computer-aided drug designing has been reported to have significant applications in almost all stages of the drug discovery pipeline in the postgenomic era. The various online resources have been categorized for the modern drug development process through the CADD process (Table 7.1).

7.9   SCOPE OF COMPUTER-AIDED DRUG DESIGNING IN THE COMING ERA

Computational methods and information technologies, coupled with statistics and chemoinformatic tools, have provided a powerful toolbox for target identification, discovery, and optimization of drug candidate molecules. According to Mallatieu,

TABLE 7.1

Softwares, Programs, or Tools in Computer-Aided Drug Designing

Sr. No.	Tools or Software	Utility of Tools or Software	Web Link	Role in CADD
1.	BLAST (Basic Local Alignment Search Tool)	A DNA and protein sequence alignment tool	https://blast.ncbi.nlm.nih.gov/Blast.cgi	Target identification and validation
2.	FASTA (Fast Alignment) Tool	A DNA and protein sequence alignment software package	http://www.ebi.ac.uk/Tools/sss/fasta/	Target identification and validation
3.	Clustal Omega	A general-purpose multiple-sequence alignment program to study evolutionary relationships	http://www.ebi.ac.uk/Tools/msa/clustalo/	Target identification and validation
4.	EMBOSS (European Molecular Biology Open Software Suite)	A free open-source software analysis package, especially developed for the needs of the molecular biology user community	http://emboss.sourceforge.net/	Target identification and validation
5.	BioEdit (Biological Editor)	A biological sequence alignment editor with multiple document interface for easy alignment and manipulation of sequences on a desktop computer	http://www.mbio.ncsu.edu/BioEdit/bioedit.html	Target identification and validation
6.	RasMol (Raster Molecule) tool	A molecular visualization program tool for DNA/RNA and protein structures	http://rasmol.org/	Structure-based drug design and target identification and validation
7.	PyMOL	Molecular visualization system for DNA/RNA and protein structures	https://www.pymol.org/	Structure-based drug design and target identification and validation
8.	Swiss-PDB Viewer	Standalone molecular visualization and modeling tool with advanced features to handle nucleic acid, proteins, and other organic molecules	http://spdbv.vital-it.ch/	Structure-based drug design and target identification and validation

(Continued)

TABLE 7.1 (*Continued*)
Softwares, Programs, or Tools in Computer-Aided Drug Designing

Sr. No.	Tools or Software	Utility of Tools or Software	Web Link	Role in CADD
9.	Discovery Studio	Advanced software focusing on modeling and simulation solutions	http://accelrys.com/products/collaborative-science/biovia-discovery-studio/	Structure-based drug design, target identification and validation, lead selection, lead optimization, and ADME studies
10.	Swiss-Modeller	Fully automated protein structure homology modeling server	https://swissmodel.expasy.org/	Structure-based drug design and target identification and validation
11.	Modeller	Standalone comparative modeling tool for 3D structures of proteins	https://salilab.org/modeller/download_installation.html	Structure-based drug design and target identification and validation
12.	PHYRE	Automatic fold recognition server for predicting the structure and function of protein sequence	http://www.sbg.bio.ic.ac.uk/phyre2/html/page.cgi?id=index	Structure-based drug design and target identification and validation
13.	PDB	An information portal to biological macromolecular structures	http://www.rcsb.org/pdb/home/home.do	Structure-based drug design and target identification and validation
14.	ISIS Draw	A chemical structure drawing program available free of cost for academic and personal use	http://mdl-isis-draw.updatestar.com/	Lead structure determination, lead optimization, and ligand-based drug design
15.	ChemDraw	A molecule editor to handle chemical molecules and is part of the ChemOffice suite of programs	https://www.cambridgesoft.com/software/overview.aspx	Lead structure determination, lead optimization, and ligand-based drug design
16.	ACD Chemsketch	Advanced chemical drawing tool available free of cost for academic use	http://www.acdlabs.com/resources/freeware/chemsketch/	Lead structure determination, lead optimization, and ligand-based drug design
17.	MarvinSketch	Advanced chemical editor for drawing chemical structures, queries, and reactions	https://www.chemaxon.com/products/marvin/marvinsketch/	Lead structure determination, lead optimization, and ligand-based drug design

(*Continued*)

TABLE 7.1 (Continued)
Softwares, Programs, or Tools in Computer-Aided Drug Designing

Sr. No.	Tools or Software	Utility of Tools or Software	Web Link	Role in CADD
18.	JME Molecular Editor	A Java applet that allows to draw/edit molecules and reactions and to depict molecules directly within an HTML page	http://www.molinspiration.com/jme/	Lead structure determination, lead optimization, and ligand-based drug design
19.	PubChem	Database containing structures and physiochemical properties of chemical compounds	https://pubchem.ncbi.nlm.nih.gov/	Lead identification, validation, and optimization
20.	CSD (Cambridge Structural Database)	Database containing experimentally determined 3D structures of potential ligand molecules	http://www.ccdc.cam.ac.uk/solutions/csd-system/components/csd/	Lead identification, validation, and optimization and ligand-based drug design
21.	ChEMBL	Chemical database of bioactive molecules with drug-like properties	https://www.ebi.ac.uk/chembl/	Lead identification, lead optimization, and ligand-based drug design
22.	OpenBabel	Open-source chemical toolbox used primarily for converting chemical file formats	http://openbabel.org/wiki/Main_Page	Lead optimization and virtual screening
23.	Molecular modeling simulation software	Molecular modeling simulation software	https://www.ncbi.nlm.nih.gov/Structure/MMDB/mmdb.shtml	Molecular docking, virtual screening, and molecular simulation
24.	ArgusLab	Molecular modeling, graphics, and drug design program	http://www.arguslab.com/arguslab.com/ArgusLab.html	Molecular docking, molecular simulation, and ligand-based drug design
25.	VegaZZ	Molecular modeling suite	http://nova.disfarm.unimi.it/cms/index.php?Software_projects:VEGA_ZZ	Molecular modeling, molecular docking, simulation, and ligand-based drug design

(Continued)

TABLE 7.1 (*Continued*)
Softwares, Programs, or Tools in Computer-Aided Drug Designing

Sr. No.	Tools or Software	Utility of Tools or Software	Web Link	Role in CADD
26.	HEX	Protein docking and molecular superposition program	http://hex.loria.fr/	Molecular docking, simulation, and ligand-based drug design
27.	QikProp	Rapid identification of ADME properties of drug candidates	http://www.schrodinger.com/	Ligand-based drug design
28.	ADMET predictor	Estimates the number of ADMET properties from query of molecular structure	http://www.simulations-plus.com/Products.aspx?grpID=1&cID=11&pID=13	ADMET property/toxicity prediction
29.	FAF-Drugs2	Subjects compounds to in silico ADMET filters	http://www.mti.univ-paris-diderot.fr/recherche/plateformes/logiciels#	ADMET property/toxicity prediction
30.	ADMET and predictive toxicology	Predicts ADMET properties	http://accelrys.com/products/discovery-studio/admet.html	ADMET property/toxicity prediction
31.	McQSAR	Uses genetic function approximation paradigm	http://users.abo.fi/mivainio/mcqsar/index.php	QSAR analysis
32.	SYBYL-X	Includes modeling and QSAR packages (CoMFA, HQSAR, and Topomer CoMFA)	http://www.certara.com/products/molmod/sybyl-x/qsar/	QSAR analysis
33.	MOLFEAT	Computes molecular fingerprints and descriptors derived from published QSAR models	http://jing.cz3.nus.edu.sg/cgi-bin/molfeat/molfeat.cgi	Molecular descriptors, prediction, and QSAR analysis
34.	Open3DQSAR	Generates chemometric analysis of molecular interaction fields	http://open3dqsar.sourceforge.net/	QSAR analysis
35.	E-Dragon	Computes molecular descriptors for structure–activity or structure–property relationship studies	http://www.vcclab.org/lab/edragon/	Molecular descriptors, prediction, and QSAR analysis

one of the senior scientists at Nutley, New Jersy, United States, working in pharmacology discovery, clearly indicates the role of future of in silico modeling in pharmaceutical discovery and development of new lead compounds. He is seriously concerned with the speed and spread of CADD. He also focused on the growth of CADD in the pharmaceutical industry for existence.

One of principal scientists at Pfizer, Lalonde, with optimistic approach, says "The ones that can successfully implement this (Insilico approaches) will probably swallowing up other companies that are not so successful, because they will keep doing it the old fashioned way and driving up the cost to astronomical levels, costs that will be very hard to justify in the marketplace. All successful companies will have to do this routinely because it's just too expensive to do it by trial and error, the way it's often been done in the past."

The growth of CADD is limited due to the lack of skilled knowledge in the industry, which requires major inputs to fulfil the challenges in the coming era.

7.10   LIMITATIONS OF COMPUTER-AIDED DRUG DESIGNING

Biological system is highly flexible, and it is not completely possible to copy and simulate the entire real biological system on computer system. Accessibility is a major problem with CADD, as many tools are not designed with a user-friendly interface in mind. The proteins and ligand molecules are highly flexible in solution because of conformation changes, and the use of a rigid structure while designing an inhibitor or drug molecule may lead to wrong result. Similarly, the effect of water molecules or other solvents should be properly incorporated into docking algorithm. Similarly, lack of concrete experimental evidence and parameters related to ADME and toxicity limits the accuracy of prediction models. Drug–drug interactions (DDIs) are of prime importance in drug safety studies, since they indicate how one drug can affect the metabolic stability of another drug. The DDIs could lead to serious health effects; therefore, predicting these effects is not only of importance but also challenging. The success rate of drug candidates getting approval for large-scale synthesis and marketing is also very less, with only 40% of drug candidates in clinical trial getting approval. Another limitation pertains to the risk that a potential, safe, and biologically active drug candidate has not been considered by predictive computational model.

7.11   FUTURE ASPECTS FOR RESEARCH BASED ON COMPUTER-AIDED DRUG DESIGNING

There is immense potential of CADD in drug development, as is evident from the successful stories of CADD application in drug discovery in recent years. Free availability of a large number of drugs such as compounds in drug databases, along with full information, supports and encourages the drug designing process. The CADD approaches can provide valuable information for target identification, validation, lead selection, small-molecular screening, and optimization. The future of CADD depends on the following:

1. Development of computational techniques for the prediction of free energies of binding and salvation.
2. Development of computational chemistry and application of new methods for carbohydrates.
3. Condensed-phase study though quantum mechanics/molecular modeling.
4. Lead structure designing against:
   a. Enzyme active sites.
   b. DNA in the minor groove.
   c. tRNA and ribozymes.
5. Novel molecular diagnostics designing based on new approaches to fluorescence.
6. Novel chemical activities such as DNA cleavage-based drug designing.
7. Small-molecule conformational properties and energetics predictions.
8. High-resolution structures of chemically modified nucleic acids or DNA: drug complexes determination by using full-distance geometry restraints, combined with high-field NMR structural determinations of nucleic acid structures.
9. How categories of ligands dock into binding sites of macromolecules.

BIBLIOGRAPHY

Accelrys Software Inc. 2013. Discovery studio modeling environment, Release 4.0. Accelrys Software, San Diego, CA.

Altschul S.F., Gish W., Miller W., Myers E.W. and Lipman D.J. 1990. Basic local alignment search tool. *J. Mol. Biol.* 215: 403–410.

Bharath E.N., Manjula S.N. and Vijaychand A. 2011. Insilico drug design tool for overcoming the innovation deficit in the drug discovery process. *Int. J. Pharm. Pharmaceut. Sci.* 3: 8–12.

Burdine L. and Kodadek T. 2004. Target identification in chemical genetics: The (often) missing link. *Chem. Biol.* 11: 593–597.

Butler M.S. 2004. The role of natural product in chemistry in drug discovery. *J. Nat. Prod.* 67: 2141–2153.

Cragg G.M. 1998. Paclitaxel (Taxol): A success story with valuable lessons for natural product drug discovery and development. *Med. Res. Rev.* 18: 315–331.

Cragg G.M. and Newman D.J. 2005. Biodiversity: A continuing source of novel drug leads. *Pure Appl. Chem.* 77: 7–24.

Der Marderosian A. and Beutler J.A. 2002. *The Review of Natural Products* (2nd ed.). Facts and Comparisons, Seattle WA, pp. 13–43.

Dimasi J.A., Hansen R.W. and Grabowski H.G. 2003. The price of innovation: New estimates of drug development costs. *J. Health Econ.* 22: 151–185.

Farnsworth N.R. 1990. The role of ethno pharmacology in drug development. In: *Ciba Foundation Symposium 154. Bioactive Compounds from Plants.* John Wiley & Sons, Chichester, UK, pp. 2–21.

Gaulton A., Bellis L.J., Bento A.P., Chambers J., Davies M. and Hersey A. 2012. Chembl: A large-scale bioactivity database for drug discovery. *Nucleic Acids Res.* 40: D1100–D1107.

Guex N. and Peitsch M.C. 1996. Swiss-PdbViewer: A fast and easy-to-use PDB viewer for macintosh and PC. *Prot. Data Bank Quat. Newslett.* 77: 7.

Harvey A.L. 2008. Natural Product in drug discovery. *Drug Discov. Today* 13: 894–901.

Irwin J., Lorber D.M., McGovern S.L., Wei B. and Shoichet B.K. 2002. Docking and drug discovery. *Comput. Nanosci. Nanotech.* 2: 50–51.

Jhoti H., Rees S. and Solari R. 2013. High-throughput screening and structure-based approaches to hit discovery: Is there a clear winner? *Expert Opin. Drug Discov.* 8: 1449–1453.

Katsila T., Spyroulias G.A., Patrinos G.P. and Matsoukas M.-T. 2016. Computational approaches in target identification and drug discovery. *Comput. Struct. Biotechnol. J.* 14: 177–184.

Kelley L.A. and Sternberg M.J.E. 2009. Protein structure prediction on the web: A case study using the Phyre server. *Nat. Protoc.* 4: 363–371.

Lavecchia A. and Di Giovanni C. 2013. Virtual screening strategies in drug discovery: A critical review. *Curr. Med. Chem.* 20: 2839–2860.

Leelananda S.P. and Lindert S. 2016. Computational methods in drug discovery. *Beilstein J. Org. Chem.* 12: 2694–2718.

Macalino S.J., Gosu V., Hong S. and Choi S. 2015. Role of computer-aided drug design in modern drug discovery. *Arch. Pharm. Res.* 38: 1686–1701.

Mandal S., Moudgil M. and Mandal S.K. 2009. Rational drug design. *Eur. J. Pharmacol.* 625: 90–100.

Mann J. 1994. *Murder, Magic, and Medicine.* Oxford University Press, New York, pp. 164–170.

McWilliam H., Li W., Uludag M., Squizzato S., Park Y.M. and Buso N. 2013. Analysis tool web services from the EMBL-EBI. *Nucleic Acids Res.* 41: 597–600.

Nag A. and Dey B. 2011. *Computer-Aided Drug Design and Delivery Systems.* The McGraw-Hill Companies, New York.

O'Boyle N.M., Banck M., James C.A., Morley C., Vandermeersch T. and Hutchison G.R. 2011. Open babel: An open chemical toolbox. *J. Chem.* 3: 33.

Oldenburg K.R. 1998. *Annual Report in Medicinal Chemistry.* Academic Press, London, UK.

Paul S.M., Mytelka D.S., Dunwiddie C.T., Persinger C.C., Munos B.H., Lindborg S.R. and Schacht A.L. 2010. How to improve R&D productivity: The pharmaceutical industry's grand challenge. *Nat. Rev. Drug Discov.* 9: 203–214.

Petrovska B.B. 2012. Historical review of medicinal plants usage. *Pharmacogn. Rev.* 6: 1–5.

Prakash P. and Gupta N. 2005. Therapeutic uses of *Ocimum sanctum* Linn (Tulsi) with a note on eugenol and its pharmacological actions: A review. *Ind. J. Physiol. Pharmacol.* 49: 125–131.

Rahman M., Karim R., Ahsan Q., Khalipha A.B.R., Chowdhury M.R. and Saifuzzaman. 2012. Use of computer in drug design and drug discovery: A review. *Int. J. Pharm. Life Sci.* 1: 1–21.

Ratti E. and Trist D. 2001. Continuing evaluation of the drug discovery process. *Pure Appl. Chem.* 73: 67–75.

Rice P., Longden I. and Bleasby A. 2000. Emboss: The European molecular biology open software suite. *Trends Genet.* 16: 276–277.

Riddle J.M. 2002. History as a tool in identifying "new" old drugs. *Adv. Exp. Med. Biol.* 505: 89–94.

Richards W.G. 1994. Computer-aided drug design. *Pure Appl. Chem.* 66: 1589–1598.

Sali A. and Blundell T.L. 1993. Comparative protein modelling by satisfaction of spatial restraints. *J. Mol. Biol.* 234: 779–815.

Schenone M., Dancik V., Wagner B.K. and Clemons P.A. 2013. Target identification and mechanism of action in chemical biology and drug discovery. *Nat. Chem. Biol.* 9: 232–240.

Schwede T., Kopp J., Guex N. and Peitsch M.C. 2003. Swiss-model: An automated protein homology-modeling server. *Nucleic Acids Res.* 31: 3381–3385.

Sliwoski G., Kothiwale S., Meiler J. and Lowe E.W. 2014. Computational methods in drug discovery. *Pharmacol. Rev.* 66: 334–395.

Sneader W. 2005. *Drug Discovery: A History.* Wiley, New York.

Song C.M., Lim S.J. and Tong J.C. 2009. Recent advances in computer-aided drug design. *Brief. Bioinform.* 10: 579–591.

Talele T.T., Khedkar S.A. and Rigby A.C. 2010. Successful applications of computer aided drug discovery: Moving drugs from concept to the clinic. *Curr. Top. Med. Chem.* 10: 127–141.

Tang Y., Zhu W., Chen K. and Jiang H. 2006. New technologies in computer-aided drug design: Toward target identification and new chemical entity discovery. *Drug Discov. Today Technol.* 3: 307–313.

Umashankar V. and Gurunathan S. 2015. Drug discovery: An appraisal. *Int. J. Pharm. Pharmaceut. Sci.* 7: 59–66.

Van Drie J.H. 2007. Computer-aided drug design: The next 20 years. *J. Comput. Aided Mol. Des.* 21: 591–601.

Verpoorte R. 1988. Exploration of a nature's chemodiversity: The role of secondary metabolites as leads in drug development. *Drug Discov. Today* 3: 232–238.

Walsh G. 2007. *Pharmaceutical Biotechnology: Concepts and Applications.* John Wiley & Sons, Chichester, UK.

Wang T., Wu M.B., Zhang R.H., Chen Z.J., Hua C., Lin J.P. and Yang L.R. 2016. Advances in computational structure-based drug design and application in drug discovery. *Curr. Top. Med. Chem.* 16: 901–916.

WEB LINKS

http://www.bioassay.de/biological-assays/introduction/biological-assays/
www.mbio.ncsu.edu/bioedit/bioedit.html
www.openrasmol.org
www.pymol.org
http://mdl-isis-draw.updatestar.com
http://www.acdlabs.com/resources/freeware/chemsketch
http://www.chemaxon.com/products/marvin/marvinsketch
http://www.molinspiration.com/jme
https://pubchem.ncbi.nlm.nih.gov
www.ccdc.cam.ac.uk/products/csd
http://mgltools.scripps.edu/downloads
http://arguslab.en.softonic.com/download
http://nova.disfarm.unimi.it/cms/index.php?Software_projects:VEGA_ZZ
http://www.schrodinger.com/
http://www.simulations-plus.com/
Products.aspx?grpID=1&cID11&pID=13
http://www.mti.univ-paris-diderot.fr/recherche/plateformes/logiciels#
http://accelrys.com/products/discovery-studio/admet.html
http://users.abo.fi/mivainio/mcqsar/index.php
http://www.certara.com/products/molmod/sybyl-x/qsar/
http://jing.cz3.nus.edu.sg/cgi-bin/molfeat/molfeat.cgi
http://open3dqsar.sourceforge.net/
http://www.vcclab.org/lab/edragon/

8 Tissue Engineering

Arushi Misra

CONTENTS

8.1 INTRODUCTION

The term *tissue engineering* was coined in 1987 at a bioengineering meet of the National Science Foundation (NSF) at Washington D.C. Tissue engineering, as stated by Robert S. Langer and Joseph P. Vacanti in their 1993 landmark research paper "is an interdisciplinary field that applies the principles of engineering and life sciences toward the development of biological substitutes that restore, maintain, or improve tissue function or a whole organ." This landmark paper consolidated different lines of research and extended preexisting ideas on tissue engineering, propelling the concepts associated with it to a higher level of understanding and awareness among researchers. According to the National Institute of Biomedical Imaging and Engineering, tissue engineering refers to the "practice of combining scaffolds, cells, and biologically active molecules into functional tissues." Although most definitions of tissue engineering encompass the fields of engineering, clinical medicine, and science, the term refers to applications that aid in the restoration, repair, or replacement of portions of or whole tissues. Tissue engineering is a multi-disciplinary field that incorporates recent advances in biomaterial science, stem cell technology, and biomimetic environments to create unique structures in the laboratory. Sometimes, the term regenerative medicine is used synonymously with tissue engineering, but the difference lies primarily in more emphasis on the use of stem cells to produce tissues in regenerative medicine.

8.2   NEED FOR TISSUE ENGINEERING

It has been observed that conventional treatments are not applicable in the case of tissues being severely diseased or lost by trauma, congenital anomaly, or cancer. In such cases, artificial tissues or organ transplantation has been the first choice for reconstruction of the devastated tissues or organs. However, these conventional surgical treatments have been marred by a shortage of donated organs, because the supply of donor organs cannot keep up with the demand. Although immunosuppressive therapy has significantly advanced in recent times, problems persist with respect to immune rejection. In addition, other available therapies such as surgical reconstruction, drug therapy, synthetic prosthesis, and medical devices are not always successful. Although artificial organs have improved in recent times due to phenomenal success in the biomedical engineering, there is still a need for better biocompatibility and biofunctionality.

With this background, man initiated the quest for finding new alternatives for better humankind. Tissue engineering is the use of a combination of cells, engineering, material science, and biochemical and physicochemical factors to improve or replace biological tissues. A unique feature of tissue engineering is to regenerate patient's own tissues and organs that are freely biocompatible and have high biofunctionality.

8.3   PROCEDURE

The general procedure of tissue engineering consists of the following steps:

- Biopsy (removal of tissue from any body part for laboratory examination)
- Isolation of specific cells
- Culturing of cells
- Selection of scaffold
- Addition of appropriate growth factors to the scaffold and cell assembly, along with mechanical stimulus, to allow it to grow in proper shape and size
- Implantation of assembly in the body of host

8.4   ELEMENTS OF TISSUE ENGINEERING

Tissue engineering contains the following elements:

1. Matrix (scaffold)
2. Cells (autologous or allogeneic)
3. Regulators

8.5   SCAFFOLD AND SCAFFOLD MATERIAL

Scaffolds are materials that are engineered to induce cellular interactions for the formation of functional tissues. They play a pivotal role in tissue regeneration and repair. Cells are implanted into a supporting device known as the scaffold, which provides support to cells to attach, grow, and differentiate. There are four major scaffolding approaches used in tissue engineering:

1. Premade porous scaffolds
2. Decellularized extracellular matrix (ECM)
3. Cells encapsulated in self-assembled hydrogels
4. Cell sheets with secreted ECM

Thus, in tissue engineering, cells are implanted into a supporting structural device called the scaffold. This facilitates the cells to remodel the scaffold into natural tissue before its implantation into the patient's body. However, sometimes, the scaffold can be placed directly into the recipient's body, where a compartment of the host's own body is utilized as a bioreactor. Figure 8.1 provides the various approaches for tissue engineering.

Scaffolds generally fulfill one or more of the following objectives:

- Allow attachment, migration, proliferation, and differentiation of cells
- Deliver and retain cells and biochemical factors
- Enable adequate nutrient and waste exchange
- Modify the behavior of the cellular phase by exerting certain mechanical and biological influences

Various synthetic and natural polymers can be used as scaffold.

1. Synthetic polymers
   a. Polylactic acid (PLA) $(C_3H_4O_2)_n$: Polylactic acid is a biodegradable, thermoplastic polyester having a melting temperature of 173°C–178°C and crystallinity of around 37%. It is popular in tissue engineering due to its biocompatibility and dissolvability in the body, where it degrades to form lactic acid, a chemical that is easily removed from the body.
   b. Polyglycolic acid (PGA) $(C_2H_2O_2)_n$: It is known since 1954 as a biodegradable, aliphatic polyester. It exhibits 45%–55% crystallinity and has a faster rate of degradation in comparison with PLA.

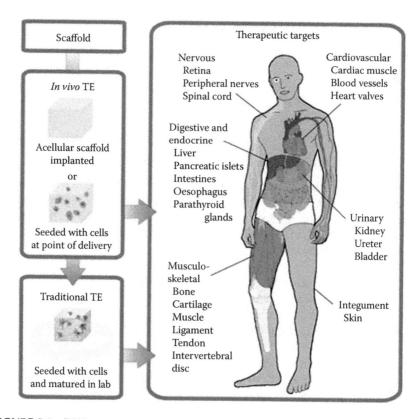

FIGURE 8.1   Different tissue engineering approaches. (Place, E.S. et al., *Chem. Soc. Rev.*, 38, 1139–1151, 2009. Reproduced by permission of The Royal Society of Chemistry.)

c. Polycaprolactone (PCL) $(C_6H_{10}O_2)_n$: Known by the IUPAC name (1,7)-polyoxepan-2-one, PCL is a biodegradable polyester with low melting point and is degraded by hydrolysis.

2. *Natural polymers*: Scaffolds may also be constructed from natural materials, particularly the different derivatives of ECM that have been studied to evaluate their ability to support cell growth. The natural polymers exhibit enhanced performance in biological systems by showing better interactions with the cells. Proteinaceous materials, such as fibrin and collagen, and polysaccharide materials, such as glycosaminoglycans (GAGs) and chitosan, have proved to be beneficial in terms of cell compatibility, barring minor issues of immunogenicity. Hyaluronic acid alone or in combination with cross-linking agents (e.g., glutaraldehyde and water-soluble carbodiimide) is considered the best choice as scaffolding material. Biomaterial scaffolds have been used for the repair and regeneration of skeletal tissues via tissue engineering strategies (Figure 8.2). Another form of scaffold under research is tissue extract from which cells

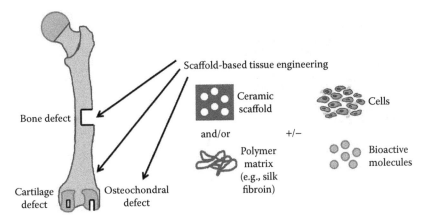

FIGURE 8.2 The concept of skeletal tissue regeneration via scaffold-based tissue engineering strategies. (Li, J.J. et al., *J. Mater. Chem. B.*, 2, 7272–7306, 2014. Reproduced by permission of The Royal Society of Chemistry.)

have been removed, and the remaining cellular remnants/extracellular matrices act as a scaffold.

3. *Nanomaterials*: Nanotechnology has been defined as the understanding and control of matter at dimensions of roughly 1–100 nanometers, where unique phenomena enable novel applications. Nanomedicine is an interdisciplinary field that amalgamates biomaterials with engineering principles for the prevention and treatment of human diseases. Carbon nanotubes (CNTs), a type of cylindrical nanostructures, have excellent mechanical properties and have been utilized to improve the mechanical and structural properties of polymer composites. The CNTs have been incorporated into polymeric scaffolds owing to their biocompatibility, resistance to biodegradation, and functionalization with biomolecules. The ECM surrounding the cells is characterized by a natural web of nanofibers and is a key factor in modulating a range of biological factors, along with playing an important role in cell–cell and cell-soluble factor interactions. Recent advances in nanobiotechnology have enabled the design and fabrication of biomimetic microenvironment that provides an analog to native ECM as well as formation of nanoscaffolds, which exhibit similarities to protein nanofibers in the ECM. Nanofibrous and nanocomposite scaffolds have been extensively researched for the regeneration of various tissues such as nerve, bone, cardiac/skeletal muscle, and blood vessels (Figure 8.3). Advantages such as incorporation of biological factors in nanofibrous scaffolds by electrospinning or self-assembly techniques can go a long way in defining the success of nanobiotechnology in tissue engineering. However, the application of nanobiotechnology in tissue engineering is still in its infancy, and a lot needs to be done in the nanoengineering field.

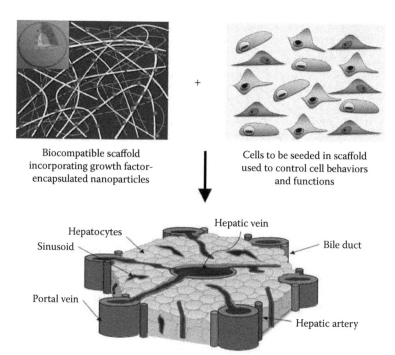

FIGURE 8.3 Nanotechnology applications in engineering complex tissues. (Reprinted with permission from Shi, J. et al., *Nano Lett.*, 10, 3223–3230, 2010. Copyright 2010 by the American Physical Society.)

8.6 GRAFTING

Grafting is a procedure in which tissues are moved from one site to another, either within an individual or between two individuals, with the help of surgery. Grafting can be done of skin, bones, ligaments, and sometimes blood vessels. In fact, skin grafting has been reported as early as 800 B.C. in India by the Hindu surgeons and developed in Europe in the eighteenth century. Skin grafting not just involves the dermal part but also strives for the restoration of all functional components, including hair follicles, sweat glands, and nerves. Bone grafting is a surgical procedure that is based on the ability of bone to regenerate completely if provided the space into which it has to grow. This technique replaces missing bone with material from patient's own body by using bone grafts as a mineral reservoir. Bone grafts promote healing of wounds and promote bone formation, since they are used as a filler and scaffold.

8.7 TYPES OF GRAFTS

1. *Autografts*: These utilize grafting material from one individual and transplant it onto another site in the same individual, for example, skin graft.

   Cells are obtained from the same individual in which they will be reimplanted. Autologous cells have the fewest problems with rejection and pathogen transmission, but in some cases, these might not be available. Recently, mesenchymal stem cells from bone marrow and fat are being used for autografting. The unique quality of these cells is that they can differentiate into a variety of tissues, including bone, cartilage, fat, and nerve. Fat is important since a large number of cells can be easily isolated from it at a rapid pace.

2. *Isograft*: Isograft, also known as syngraft, is the graft taken from one individual and transplanted in another individual having the same genetic constitution, that is, a graft between highly inbred strains or identical twins.

3. *Allograft*: It is also termed homeostatic. Graft is harvested from one individual and placed on genetically nonidentical member of the same species. Allografts are recognized by the immune system of an individual and could be rejected after transplantation. Most of the human-to-human grafts are allografts. Examples are spinal transfusion, heart valves, liver transplant, kidney transplant, bone marrow transplant, and so on.

4. *Xenograft*: It is a graft taken from a species other than human. Particularly, animal cells have been used extensively in experiments aimed at the construction of cardiovascular implants, for example, transplantation from animals to human beings.

8.8   SYNTHESIS OF SCAFFOLDS

Numerous methods are available for preparing porous structures that can be utilized as tissue engineering scaffolds. Some of these are mentioned as follows:

8.8.1   NANOFIBERS

Molecular self-assembly is a method for creation of biomaterials having properties similar in scale and chemistry to that of the natural *in vivo* ECM. Self-assembly of synthetic or natural molecule yields nanoscale fibers, known as nanofibers. Although a complicated and elaborate process, self-assembly shows several advantages such as production of a very thin nanofiber with very thin diameter, negligible use of organic solvent, and reduction in the cytotoxicity, because it is carried out in aqueous solution containing salt or physiological media.

8.8.2   PHASE SEPARATION

Phase separation refers to the dissolution of a biodegradable synthetic polymer in molten phenol or naphthalene, followed by addition of biologically active molecules such as alkaline phosphatase to the solution. The lowering of the temperature separates a liquid–liquid phase, which is then quenched to form a two-phase solid. The above-mentioned steps are followed by removal of the solvent by sublimation, which gives a porous scaffold having bioactive molecules incorporated in the structure. The primary advantage of this technique is its easy combination with other fabrication technology to design three-dimensional structures with controlled pore morphology.

The phase separation technique can also be integrated with rapid prototyping (RP) for creation of nanofibrous scaffolds that find use in tissue engineering.

8.8.3 SOLVENT CASTING AND PARTICULATE LEACHING

Among the various scaffolding techniques, solvent casting and particulate leaching (SCPL) is known for its simplicity; cost-effectiveness; possibility of producing scaffolds with a good pore interconnectivity degree; and controlled composition, pore size, and porosity. This technique involves producing a solution of PLA in chloroform and thereafter adding particles of a particular thickness to produce a uniform suspension. The added particles could be salt such as sodium chloride, gelatin spheres, crystals of saccharose, or paraffin spheres. After the polymer solution has been cast, evaporation of the solvent leaves behind a polymer matrix having salt particles embedded throughout. The composite is then immersed in a bath of liquid (water is suitable for sodium chloride), where the salt leaches out to produce a porous structure. However, the major drawbacks of this method include the following:

1. A small thickness range is obtained.
2. The organic solvents must be wholly removed to avoid any possible damage to the cells seeded on the scaffold.
3. The toxic solvent can denature the protein and may affect other solvents.
4. Retention of toxicity by the scaffolds designed by this technique.

8.8.4 GAS FOAMING

The gas foaming technique uses carbon dioxide gas at high pressure for the synthesis of highly porous scaffolds, thus overcoming the need to use organic solvents, solid particles, and high temperature. In this technique, disc-shaped structures are made of the desired polymers by compression molding, and these are then placed in a chamber, where they are exposed to high-pressure carbon dioxide (800 psi) for several days to saturate the polymer with gas. The pressure inside the chamber is gradually restored to atmospheric levels, resulting in the formation of pores by the carbon dioxide molecules that abandon the polymer, resulting in a decrease in polymer density and leading to the formation of a sponge-like structure. The use of porogens such as sugar, salts, and wax controls the porosity of the scaffolds.

8.8.5 FREEZE DRYING

This technique is based on the principle of sublimation and is used for the fabrication of porous scaffolds. In this technique, first, a synthetic polymer is dissolved into a suitable solvent to form a solution of specific concentration. For example, PGA is dissolved in benzene or glacial acetic acid. Thereafter, the solution is frozen, and the solvent is removed by lyophilization under high-vacuum condition, which leads to the synthesis of the scaffold having properties such as high porosity and interconnectivity. Although the pore size can be controlled by the freezing rate and pH, a rapid freezing rate produces smaller pores. A smaller pore size and long processing time are the major drawbacks of this technique, whereas the main advantage is that this technique neither requires high temperature nor separate leaching step.

8.8.6 ELECTROSPINNING

Electrospinning is a highly versatile technique that can be used to produce fibers with diameters in the nanoscale or microscale range. In a typical electrospinning setup, a solution is diverted through a spinneret and a high-intensity electric field is applied to the tip. The buildup of electrostatic current within the charged solution causes it to eject a stream of polymer solution or melt, which gives rise to polymer fiber after drying or solidification. A number of polymers such as chitosan, collagen, gelatin, and silk fibroin have been used for electrospinning. The main advantages of this technique are its simplicity and production of ultrafine particles with special orientation, high surface area, high aspect ratio, and control over pore geometry.

8.8.7 COMPUTER-AIDED DESIGN AND COMPUTER-ASSISTED MANUFACTURING TECHNOLOGIES

The rapid development of bioinformatics and computational biology has led to the use of computer-aided design (CAD)-based manufacturing technologies for the fabrication of scaffolds. The use of CAD with computer-assisted manufacturing (CAM) enables researchers to reproduce complex scaffold structures without any requirement of molds. The application of computer-aided technologies in tissue engineering has led to the development of a new field known as computer-aided tissue engineering (CATE). The CATE has been applied for designing and fabrication of tissue scaffolds, which facilitates the incorporation of biomimetic and biological features into the scaffold design.

8.9 ASSEMBLY AND BIOPRINTING

The ultimate aim of tissue engineering is the assembly of different types of cells into one functional tissue or organ. However, the engineered tissues generally lack an initial blood supply, which makes it difficult for any implanted cell to obtain nutrients and oxygen, and this can lead to improper functioning. Tissue assembly from cellular elements could be achieved by numerous approaches such as self-assembly, bioprinting, dielectrophoretic patterning, micromolding, magnetic levitation, and robots.

Self-assembly, the autonomous organization of components, from an initial pattern into the final state, without external intervention, plays an important role here. It has been recognized by the tissue engineering researchers that the success of engineering functional living structures will depend on a thorough understanding of the principles of cellular self-assembly and our ability to use them.

Three-dimensional (3D) printing, first introduced by Charles W. Hull in 1986, creates 3D structures by adding layer by layer of material and is being increasingly utilized in tissue engineering. The material can be ceramic, metal, plastic, or polymers (synthetic or natural). This technique is also referred to as additive manufacturing (AM) or RP. Engineering of 3D living structures using bioprinting is being deeply looked into by tissue engineering researchers. This can be done by the following two approaches:

1. The first approach relies on the use of inkjet printing, wherein either individual cells or small clusters are printed. This method is cheap, quick, and quite versatile, but it is marred by considerable damage to cells and difficulty in assuring high cell density needed for the fabrication of solid organ structures.

2. In the second approach, mechanical extruders are used to place *bioink* particles into a supporting environment, the *biopaper*, per the computer-generated templates that are consistent with the topology of the desired biological structure. The organoids are formed by the postprinting fusion of the bioink particles and the arrangement of cells within the bioink particles. This technique, though incurring high cost, is more advantageous since the cells in the 3D tissue fragments are in a more relevant arrangement, with adhesive contacts with their neighboring cells, which may assist in the transmission of important molecular signals.

8.10   SUCCESS STORIES OF TISSUE ENGINEERING

The rapid strides in tissue engineering and associated technologies such as biomaterial science, biochemistry, molecular medicine, nanotechnology, and biomedical engineering have led to the increasing role of tissue engineering in tissue and organ regeneration in the field of medicine. Tissue engineering may offer new alternatives for repair or replacement of diseased or deteriorated organs. Thus, it has major advantages over traditional organ transplantation and bypasses the problem of organ shortage.

8.10.1   Cardiovascular Tissue Engineering

Cardiac tissue engineering aims to mimic the native environment of the cardiac tissue of the host by construction of new tissue that facilitates the assimilation of the bioengineered, *donor* tissue by the host milieu. The main objective of cardiac tissue engineering is to effectively repair or replace the injured heart muscle. An ideal cardiac tissue construct should have the following properties that are similar to those of the native myocardium:

1. It should not be immunologically rejected by the host or elicit an inflammatory response *in vivo*; that is, it should have biocompatibility with the host tissue.
2. It should remain viable before and after delivery and/or implantation.
3. It should be biomimetic.
4. It should facilitate cell–cell adhesion, generate force comparable with the native tissue during contraction, and conduct electrical signals.

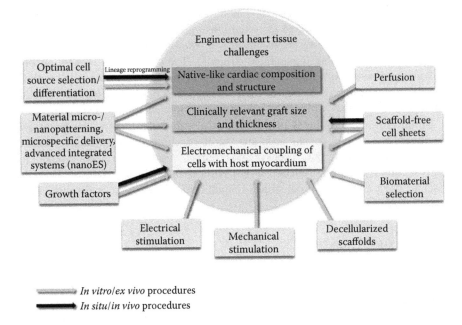

In vitro/ex vivo procedures
In situ/in vivo procedures

FIGURE 8.4 Integrative approach of different cardiac tissue engineering strategies and technologies in addressing the main cardiac tissue engineering challenges. (Georgiadis, V. et al., *Integr. Biol.*, 6, 111–126, 2014. Reproduced by permission of The Royal Society of Chemistry.)

Cardiovascular tissue engineering aims for the engineering of 3D cardiac tissue. The last three decades have witnessed massive leaps toward restoring the physiological functions of the heart through effective delivery of cell implants into the organ, initially through *naked cell* injections and thereafter by cardiac tissue engineering. Intricate delivery systems and priming mechanisms have been utilized to fulfill this aim, which includes the use of scaffolds and bioreactors and manipulations of cells and support systems. Various strategies employed in heart tissue engineering are unique in terms of their aim and design and converge at the main *checkpoints* that define an optimal construct or implant (Figure 8.4).

Vunjak-Novakovic et al. (2011) identified six different cardiac engineering approaches:

1. Mechanical stimulation of cells in hydrogels
2. Electrical stimulation of cells contained in porous scaffolds
3. Culturing of cells on perfused channeled scaffolds
4. Decellularization of the native heart and repopulation with donor cells
5. *In situ* cell delivery via injectable hydrogels
6. Culture of cell sheets that are scaffold-free

The intended cell sources are isolated and cultivated from the biopsy (vein, umbilical cord, or artery). Simultaneously, the components of the fibrin are also removed from a patient's blood sample. In an injection molding procedure, a tricuspid or right atrio-ventricular valve is synthesized from the fibrin gel matrix and the myofibroblasts. The synthesized heart valve is subsequently *conditioned* in a bioreactor system for the formation of a stable implantable structure. The second step makes use of a dynamic cultivation in a bioreactor for coating of the surface of the valve with an endothelial cell lining. The end of the manufacturing process yields a completely autologous heart valve, which can be subsequently implanted into the patient.

Despite the rapid progress in this field, the restoration or replacement of cardiac muscles is a cumbersome task, and therefore, clinically significant contractile improvements are quite rare. There are no reports of a safe and efficient protocol for the restoration of cardiac function in any clinical study. Poor electrochemical and vascular integration of cell constructs remain critical areas that lead to low functional gains.

8.10.2 Nerve Engineering

Repair of neuronal tissues and regeneration strategies have received a lot of attention, because these are directly concerned with the quality of the patient's life. Numerous strategies such as implantation of autografts, allografts, and xenografts have been employed to increase the prospects of axonal regeneration and functional recovery. Autograft is by far considered the best choice, wherein a segment of nerves is removed from other body parts. However, this approach has demerits such as limited supply of available nerve graft material and permanent loss of nerve function of the donor. Allograft and xenografts have also been frequently used but are limited due to the need of immunosuppressive therapy and varying success rates. The artificial nerve graft that avoids the problems of availability and immune rejection could be a promising alternative for extending the length over which nerve can successfully regenerate. The materials used must be altered to make them cell friendly. Variations in chemical and physical properties of artificial material are attractive, allowing alterations in geometric configuration, biocompatibility, porosity, degradation, electrical conductivity, and mechanical strength. Fabrication of biodegradable nerve guidance channels based on chitin and chitosan has been successfully accomplished for improvements in nerve tissue engineering. Wang et al. (2005) bridged a 30 mm sciatic nerve defect in a large animal model by using an artificial nerve graft of chitosan–PGA blend.

8.10.3 Skin Engineering

The skin is the largest organ of the human body. It is important as it provides a physical barrier to the external environment, thermal regulation, and retention of normal hydration. Skin consists of three layers, viz. epidermis, dermis, and hypodermis, and represents approximately one-tenth of the body mass that also acts as a barrier to pathogens and trauma. Skin destruction caused by injury, burns, venous ulcers, and diabetic ulcers occasionally induces life-threatening situations. This led to the development of skin grafting technique, which may be from the same human subject (autograft), other humans (allografts), or animals (xenografts) or by using membranes fabricated from natural or synthetic polymers. Autografting in skin is rarely resorted to because of the limited availability of the donor tissue. Owing to this, wounds are usually covered with allografts that are harvested from consenting donors after death and stored in skin banks in frozen condition. Allografts often serve as temporary covering, because they are rejected by the host's immune system within a short duration. The growth factors released from such grafts have a positive effect on wound healing till the time an autograft can be placed onto the wound. Consequently, intensive research led to skin becoming the first tissue to be successfully engineered in the laboratory. This was accomplished in two steps: the first involves the development of biodegradable matrix that simulates the dermis, and the second step leads to the development of keratinocyte culture, leading to live cultured skin products.

Skin-specific stem cells have opened new avenues in skin engineering. Stem cells, due to their unique ability to differentiate into various types of tissue by asymmetric replication, have the potential to create those components of the skin that are not found in the engineered skin. Spindle-shaped mesenchymal stem cells, also termed mesenchymal stromal cells, have significantly contributed to cellular therapy and skin tissue engineering. Adipose stem cells (ASCs) highly contribute to skin repair by improving neoepidermis formation and skin aging by accelerating angiogenesis and multiplication of fibroblasts. A mixture of ASCs and fibroblasts improves the epidermal morphogenesis of tissue-engineered skin. Properties such as the ability of ASCs to grow on porous biomaterial and to differentiate into endothelial, fibrovascular, and epithelial tissues for skin repair point toward the excellent capability of ASCs as a cellular source for skin tissue engineering.

8.10.4 Bone Engineering

Bone is a highly dynamic and diverse tissue that represents the foundation for our bodily locomotion and performs various functions such as providing the load-bearing capacity to our skeleton, housing the elements required for hematopoiesis, and maintaining the homeostasis of key electrolytes via calcium and phosphate ion storage. Bone is the calcified connective tissue mainly consisting of a natural organic mineral composed of collagen type I and diverse forms of calcium phosphate. There is an outer layer of low porosity (10%–30%) and high mechanical strength and an inner layer of higher porosity (50%–90%), with both layers being highly vascularized. Bones possess good capacity for regeneration, remodeling, and repair in response to injury, but bone grafting is resorted to when the required bone regeneration exceeds the natural regeneration process for self-healing. However, autologous and allogeneic transplantations by using autografts and allografts for bone repair and regeneration have numerous shortcomings, limitations, and complications. The tissue engineering approach for bone defect repair is perceived as a better approach as compared with conventional approaches such as grafting, since the repair process may proceed with the patient's own tissue by the time the regeneration is complete. Bone tissue engineering is a dynamic and highly complex process that is initiated by the migration and recruitment of osteoprogenitor cells, their multiplication, differentiation, and matrix formation and culminates in remodeling of the bone. An ideal scaffold for use in bone engineering should have specific features such as biocompatibility, bioresorbability, optimum pore size (200–350 μm), and mechanical properties matching the host bone properties. A number of biomolecules that can aid bone tissue engineering such as proteins and growth factors (transforming growth factor [TGF]-β, bone morphogenetic protein [BMP], insulin-like growth factor [IGF], fibroblast growth factor [FGF], and vascular endothelial growth factor [VEGF]) are delivered via scaffolds. Among the various fabrication techniques available, solid freeform fabrication (SFF)-based techniques have been the most widely used for fabrication of 3D interconnected porous scaffolds. Scaffolds made with different materials such as calcium phosphate-based bioactive ceramics, bioglass, composites, polymers, and metals (titanium and tantalum) have been tested under *in vitro* and *in vivo* conditions. Currently, third-generation biomaterials are being fabricated to stimulate regeneration of bones by using tissue engineering and *in situ* tissue regeneration methods. However, despite rapid strides in bone engineering, several key challenges such as biomechanical strength and biocompatibility of polymer scaffolds, biodegradation, limited bioactivity, and release of metallic ions in metallic scaffolds, toughness and reproducible manufacturing techniques for ceramic scaffolds still remain and are being worked into.

8.10.5 Liver Engineering

The liver is the largest organ in the human body, with complex functions such as glucose and lipid metabolism, homeostasis, detoxification, secretion of bile, and production of serum proteins. Liver function decreases on destruction of hepatocytes, resulting in liver failure. In this condition, patients generally go for liver transplantation, which is marred by high cost, a shortage of donors, and a relatively risky operation. Thus, there is an urgent need of alternatives to fill the gap between organ availability and demand and to reduce the health burden. Bioengineering can play a major role in the liver regeneration field and the treatment of liver diseases by providing decellularization–recellularization methods for maintaining the anatomical structure of organs and allowing hepatic cells to form physiological contacts. Rapid strides in the development of new biomaterials platforms that can collectively modulate host cell recruitment and material-resident cell function, along with engraftment and release of implanted cell types, have shown great promise in liver tissue engineering approaches. Three-dimensional bioprinting, an advanced technique that uses biofabrication techniques to build 3D spheroids by a layer-by-layer approach using cell suspensions, enables the accurate construction of complex parenchymal organ structures, including a complex vascular tree network. The ability of adult, embryonic, and induced pluripotent stem cells to differentiate into liver cells has been exhaustively studied. Several researchers have studied the differentiation of embryonic stem cells into hepatocyte-like cells that can be used in bioartificial liver devices, which provide time for a diseased liver to recover, without the potential zoonoses associated with using porcine hepatocytes. However, issues such as immunocompatibility, possible teratoma formation by nondifferentiated cells, and ethical considerations remain. With rapid progress in tissue engineering and bioprinting techniques, the development of artificial livers seems quite feasible in the coming years.

8.11 CONCLUSION

Tissue engineering has emerged as a vibrant industry, with rapid development over the past 30 years and the groundbreaking research originating in both laboratory and clinical settings. The coming decades are likely to witness major drive toward integrating sophisticated architectures and cellular signals that will help in accelerating repair and integration to give constructs that would accommodate the three dimensions of functional systems and will combine with real-time responses over extended life spans. Recent advances suggest that engineered tissues may have greater clinical applications in the future and are likely to represent an excellent therapeutic option for those who would benefit from the life-extending benefits of tissue replacement or repair. Future biomaterials will be designed at the molecular level based on new concepts and principles to be utilized *in situ* by cells and tissues to accelerate the reconstruction and regeneration of tissues and organs.

BIBLIOGRAPHY

Amini A.R., Laurencin C.T. and Nukavarapu S.P. 2012. Bone tissue engineering: Recent advances and challenges. *Crit. Rev. Biomed. Eng.* 40: 363–408.

Atala A. 1999. Engineering tissues and organs. *Curr. Opin. Urol.* 9: 517–526.

Auger F.A., Berthod F., Moulin V., Pouliot R. and Germain L. 2004. Tissue-engineered skin substitutes: From in vitro constructs to in vivo applications. *Biotechnol. Appl. Biochem.* 39: 263–275.

Bhatia S.N., Underhill G.H., Zaret K.S. and Fox I.J. 2014. Cell and tissue engineering for liver disease. *Sci. Transl. Med.* 6: 245sr2.

Boland E.D., Espy P.G. and Bowlin G.L. 2004. Tissue engineering scaffolds. In: Wenk G.E. and Bowlin G.L. (Eds.) *Encyclopedia of Biomaterials and Biomedical Engineering.* Marcel Dekker, Richmond, VA, pp. 1633–1635.

Bose S., Roy M. and Bandyopadhyay A. 2012. Recent advances in bone tissue engineering scaffolds. *Trends Biotechnol.* 30: 546–554.

Bottcher-Haberzeth S., Biedermann T. and Reichmann E. 2010. Tissue engineering of skin. *Burns* 36: 450–460.

Castells-Sala C., Alemany-Ribes M., Fernandez-Muiños T., Recha-Sancho L., Lopez-Chicon P., Aloy-Reverté C., Caballero-Camino J., Márquez-Gil A. and Semino C.E. 2013. Current applications of tissue engineering in biomedicine. *J. Biochip. Tissue Chip.* S2: 004.

Chan B.P. and Leong K.W. 2008. Scaffolding in tissue engineering: General approaches and tissue-specific considerations. *Eur. Spine J.* 17(Suppl 4): 467–479.

Chen M., Przyborowski M. and Berthiaume F. 2009. Stem cells for skin tissue engineering and wound healing. *Crit. Rev. Biomed. Eng.* 37(4–5): 399–421.

Cui F.Z. and Mikos A.G. 2014. Important topics in the future of tissue engineering: Comments from the participants of the 5th International Conference on Tissue Engineering at Kos, Greece. *Regener. Biomat.* 1: 103–106.

Dhandayuthapani B., Yoshida Y., Maekawa T. and Kumar D.S. 2011. Polymeric scaffolds in tissue engineering application: A review. *Int. J. Polym Sci.* 2011: 19p.

Dhawan A., Puppi J., Hughes R.D. and Mitry R.R. 2010. Human hepatocyte transplantation: Current experience and future challenges. *Nat. Rev. Gastroenterol. Hepatol.* 7: 288–298.

Fuchs J.R., Nasseri B.A. and Vacanti J.P. 2001. Tissue engineering: A 21st century solution to surgical reconstruction. *Annals Thorac. Surg.* 72: 577–591.

Georgiadis V., Knight R.A., Jayasinghe S.N. and Stephanou A. 2014. Cardiac tissue engineering: Renewing the arsenal for the battle against heart disease. *Integr. Biol.* 6: 111–126.

Gu B.K., Choi D.J., Park S.J., Kim M.S., Kang C.M. and Kim C.H. 2016. 3-dimensional bioprinting for tissue engineering applications. *Biomaterials Res.* 20: 12.

Gunatillake P.A. and Adhikari R. 2003. Biodegradable synthetic polymers for tissue engineering. *Eur. Cell Mater.* 5: 1–16.

Hansbrough J.F. and Franco E.S. 1998. Skin replacements. *Clin. Plast. Surg.* 25: 407–423.

Henkel J., Woodruff M.A., Epari D.R., Steck R., Glatt V., Dickinson I.C., Choong P.F.M., Schuetz M.A. and Hutmacher D.W. 2013. Bone regeneration based on tissue engineering conceptions- a 21st Century Perspective. *Bone Res.* 1: 216–248.

Horch R.E. 2006. Future perspectives in tissue engineering: "Tissue engineering" review series. *J. Cell. Mol. Med.* 10: 4–6.

Horch R.E., Kneser U., Polykandriotis E., Schmidt V.J., Sun J. and Arkudas A. 2012. Tissue engineering and regenerative medicine—where do we stand? *J. Cell. Mol. Med.* 16: 1157–1165.

Hutmacher D.W. 2001. Scaffold design and fabrication technologies for engineering tissues-state of the art and future perspectives. *J. Biomater. Sci. Polym. Ed.* 12: 107–124.

Ikada Y. 2006. Challenges in tissue engineering. *J. R. Soc. Interface* 3: 589–601.

Ikada Y. 2006. *Tissue Engineering: Fundamentals and Applications.* Academic Press, San Diego, CA.

Jakab K., Marga F., Norotte C., Murphy K., Vunjak-Novakovic G. and Forgacs G. 2010. Tissue engineering by self-assembly and bio-printing of living cells. *Biofabrication* 2(2): 022001.

Jawad H., Ali N.N., Lyon A.R., Chen Q.Z., Harding S.E. and Boccaccini A.R. 2007. Myocardial tissue engineering: A review. *J. Tissue Eng. Regen. Med.* 1(5): 327–342.

Khan Y., Yaszemski M.J., Mikos A.G. and Laurencin C.T. 2008. Tissue engineering of bone: Material and matrix considerations. *J. Bone Joint Surg. Am.* 90: 36–42.

Kim B.S. and Mooney D.J. 1998. Development of biocompatible synthetic extracellular matrices for tissue engineering. *Trends Biotechnol.* 16: 224–230.

Kim E.-S., Ahn E.H., Dvir T. and Kim D.-H. 2014. Emerging nanotechnology approaches in tissue engineering and regenerative medicine. *Intern. J. Nanomed.* 9(Suppl 1): 1–5.

Knight M.A. and Evans G.R. 2004. Tissue engineering: Progress and challenges. *Plast. Reconstr. Surg.* 114: 26E–37E.

Langer R. and Vacanti J.P. 1993. Tissue engineering. *Science* 260: 920–926.

Lanza R. and Vacanti J. 2007. *Principles of Tissue Engineering.* Elsevier Academic Press, Boston, MA.

Lee S.Y., Kim H.J. and Choi D. 2015. Cell sources, liver support systems and liver tissue engineering: Alternatives to liver transplantation. *Int. J. Stem Cells* 8: 36–47.

Li J.J., Kaplan D.L. and Zreiqat H. 2014. Scaffold based regeneration of skeletal tissues to meet clinical challenges. *J. Mater. Chem. B.* 2: 7272–7306.

Lichte P., Pape H.C., Pufe T., Kobbe P. and Fischer H. 2011. Scaffolds for bone healing: Concepts, materials and evidence. *Injury* 42: 569–573.

Lu T., Li Y. and Chen T. 2013. Techniques for fabrication and construction of three-dimensional scaffolds for tissue engineering. *Int. J. Nanomedicine* 8: 337–350.

Ma P.X. 2004. Scaffolds for tissue fabrication. *Mater. Today* 7: 30–40.

Ma P.X. and Elisseeff J. 2005. *Scaffolding in Tissue Engineering.* CRC Press, Boca Raton, FL.

MacArthur B.D. and Oreffo R.O. 2005. Bridging the gap. *Nature* 433: 19.

MacNeil S. 2207. Progress and opportunities for tissue-engineered skin. *Nature* 445: 874–880.

Mikos A.G. and Temenoff J.S. 2000. Formation of highly porous biodegradable scaffolds for tissue engineering. *Electron. J. Biotechnol.* 3: 114–119.

Mironov V., Boland T., Trusk T., Forgacs G. and Markwald R.R. 2003. Organ printing: Computer-aided jet-based 3D tissue engineering. *Trends Biotechnol.* 21: 157–161.

Muschler G.F., Nakamoto C. and Griffith L.G. 2004. Engineering principles of clinical cell-based tissue engineering. *J. Bone Joint Surg. Am.* 86: 1541–1558.

Nomi M., Atala A., Coppi P.D. and Soker S. 2002. Principals of neovascularization for tissue engineering. *Mol. Aspects Med.* 23: 463–483.

Olson J.L., Atala A. and Yoo J.J. 2011. Tissue engineering: Current strategies and future directions. *Chonnam Med. J.* 47: 1–13.

Oshima H., Rochat A., Kedzia C., Kobayashi K. and Barrandon Y. 2001. Morphogenesis and renewal of hair follicles from adult multipotent stem cells. *Cell* 104: 233–245.

Pellegrini G., Bondanza S., Guerra L. and De Luca M. 1998. Cultivation of human keratinocyte stem cells: Current and future clinical applications. *Med. Biol. Eng. Comput.* 36: 778–790.

Pham Q.P., Sharma U. and Mikos A.G. 2006. Electrospinning of polymeric nanofibers for tissue engineering applications: A review. *Tissue Eng.* 12: 1197–1211.

Place E.S., George J.H., Williams C.K. and Stevens M.M. 2009. Synthetic polymer scaffolds for tissue engineering. *Chem. Soc. Rev.* 38: 1139–1151.

Sachlos E. and Czernuszka J.T. 2003. Making tissue engineering scaffolds work. Review: The application of solid freeform fabrication technology to the production of tissue engineering scaffolds. *Eur. Cell Mater.* 5: 29–40.

Shevchenko R.V., James S.L. and James S.E. 2010. A review of tissue-engineered skin bioconstructs available for skin reconstruction. *J. R. Soc. Interface* 7: 229–258.

Shi J., Votruba A.R., Farokhzad O.C. and Langer R. 2010. Nanotechnology in drug delivery and tissue engineering: From discovery to applications. *Nano Lett.* 10: 3223–3230.

Srivastava S. and Bhargava A. 2012. Nanomedicine: The next generation medicine. In: Bhargava A. and Srivastava S. (Eds.) *Biotechnology: New Ideas, New Developments.* Nova Science Publishers, Hauppauge, NY, pp. 115–136.

Subramanian A., Krishnan U.M. and Sethuraman S. 2009. Development of biomaterial scaffold for nerve tissue engineering: Biomaterial mediated neural regeneration. *J. Biomed. Sci.* 16: 108.

Sun B.K., Siprashvili Z. and Khavari P.K. 2014. Advances in skin grafting and treatment of cutaneous wounds. *Science* 346: 941–945.

Sun W., Darling A., Starly B. and Nam J. 2004. Computer-aided tissue engineering: Overview, scope and challenges. *Biotechnol. Appl. Biochem.* 39: 29–47.

Supp D.M. and Boyce S.T. 2005. Engineered skin substitutes: Practices and potentials. *Clin. Dermatol.* 23: 403–412.

Tsang V.L. and Bhatia S.N. 2007. Fabrication of three-dimensional tissues. *Adv. Biochem. Eng. Biotechnol.* 103: 189–205.

Vacanti C.A. 2006. History of tissue engineering and a glimpse into its future. *Tissue Eng.* 12: 1137–1142.

Vacanti C.A. and Vacanti J.P. 2000. The science of tissue engineering. *Orthop. Clin. North Am.* 31: 351–356.

Vunjak-Novakovic G. and Kaplan D.L. 2006. Tissue engineering: The next generation. *Tissue Eng.* 12: 3261–3263.

Vunjak-Novakovic G., Lui K.O., Tandon N. and Chien K.R. 2011. Bioengineering heart muscle: A paradigm for regenerative medicine. *Annu. Rev. Biomed. Eng.* 13: 245–267.

Wang X., Hu W., Cao Y., Yao J., Wu J. and Gu X. 2005. Dog sciatic nerve regeneration across a 30-mm defect bridged by a chitosan/PGA artificial nerve graft. *Brain* 128: 1897–1910.

Whitesides G.M. and Boncheva M. 2002. Beyond molecules: Self-assembly of mesoscopic and macroscopic components. *Proc. Natl. Acad. Sci. (USA)* 99: 4769–4774.

Whitesides G.M. and Grzybowski B. 2002. Self-assembly at all scales. *Science* 295: 2418–2421.

Yannas I.V. and Burke J.F. 1980. Design of an artificial skin. I. Basic design principles. *J. Biomed. Mater. Res.* 14: 65–81.

Yu X. and Bellamkonda R.V. 2003. Tissue-engineered scaffolds are effective alternatives to autografts for bridging peripheral nerve gaps. *Tissue Eng.* 9: 421–430.

9 Phytomining
Principles and Applications

Meenakshi Bhargava and Vandana Singh

CONTENTS

9.1 INTRODUCTION

Mining has been at the forefront of economic activity in human development. Commercial mining is carried out from ores that have a high concentration of target metals and requires huge capital investment. The ores occur in small localized areas and are being depleted as a result of population increase, expanding economies, and misdirected industrialization. There are four main mining methods, viz. underground, surface (pit), placer, and *in situ* mining. The choice of the method depends on the type of mineral that is to be obtained, its location, and its economic feasibility. Each mining method has its own merits and demerits, with varying degrees of environmental impact.

   1. *Underground mining*: It refers to various underground mining techniques that are used to remove minerals and ores containing metals such as iron, lead, copper, zinc, nickel, platinum, gold, and silver. Underground

mining is also used for excavating ores of gems such as diamonds. This technique is more expensive and is often used to reach deposits that are deeply placed.

2. *Surface mining*: This process involves the removal of soil and rock overlying the mineral deposit, referred to as overburden. This technique is essentially used when deposits of commercially important minerals or rock are found close to the surface or material is structurally unsuitable for heavy handling or tunneling (as would be in the case of less valuable deposits).

3. *Placer mining*: Placer deposits are concentrations of elements that are formed when minerals are washed, by weather or flooding, downslope into streams. Placer mining refers to the collection of mining methods that use water to separate ores (mostly of precious metals such as gold) from the surrounding alluvial or placer deposits. The technique, sometimes referred to as environmentally destructive, is used to take out valuable metals and gemstones from sediments in river channels, beach sands, or similar environments.

4. In situ *mining*: This technique involves the recovery of minerals in the ground by dissolving them and pumping the resulting solution to the surface, where the minerals can be recovered. *In situ* mining is primarily used for mining of copper and uranium.

In the past, commercial mining has been based solely on economic considerations, with scarce regard to ramifications on the environment. However, with changing times, environmental awareness and management have also become the important priorities for governments, researchers, and policy makers. A very large area of the earth is covered by low-grade ores associated with ultramafic deposits, in which the percentage of metal is well below the metal content required to be economically extracted and smelted by conventional techniques. Weathering of ultramafic rocks produces the ultramafic or serpentine soils, which are characterized by low ratio of Ca to Mg in the exchangeable cation and soil solution, less amount of organic matter, and relatively large amounts of heavy metals such as cobalt, nickel, iron, chromium, magnesium, titanium, and other such types. These soils, constituting approximately 1% of the earth's land surface, are distributed throughout the world and support endemic flora that has constitutive and adaptive mechanisms for tolerating high concentration of metals.

9.2 HEAVY METALS AND THEIR IMPORTANCE TO PLANTS AND ANIMALS

Heavy metals, the ubiquitous environmental contaminants, are members of a loosely defined group of elements that have a specific gravity of more than 5 g/cm^3 in their standard state. Per the above-mentioned criterion, 53 elements are regarded as heavy metals. Some of these heavy metals are of importance to living forms, whereas others are toxic. Heavy metals such as iron (Fe), manganese (Mn), zinc (Zn), copper (Cu), cobalt (Co), and molybdenum (Mo) are essential for the growth of life forms, whereas others have a single function, such as vanadium (V) in some peroxidases and nitrogenases and nickel (Ni) in hydrogenases. Some of the heavy metals such as cadmium (Cd), lead (Pb), uranium (U), thallium (Tl), chromium (Cr), silver (Ag), and mercury (Hg) are of no utility to the living forms and have varying degrees of toxicity.

Metals play a variety of roles in all living organisms and are important, since they are the active centers of many enzymes. These have been utilized for catalyzing key reactions in the living forms and for maintaining protein structure. Metals are required in small amounts for normal cell metabolism and play a pivotal role as structural elements, stabilizers of biological structures, components of control mechanisms, and activators or components of redox systems. Some of the metals are essential for human body, and their deficiency results in disordered biological functions, whereas some metals such as mercury and cadmium have no known biological function. Table 9.1 provides information on some of the essential roles of metals in plant systems.

TABLE 9.1
Importance of Heavy Metals for Plants

Metal	Beneficial Effects of Heavy Metals	Reference
Cu	Important role in CO_2 assimilation and ATP synthesis.	Thomas et al. (1998)
	Component of plastocyanin and cytochrome oxidase.	Demirevska-kepova et al. (2004)
Fe	Synthesis of chlorophyll.	Miller et al. (1995); Spiller et al. (1982)
	Component of cytochromes.	Soetan et al. (2010)
Zn	Synthesis of cytochrome.	Tisdale et al. (1984)
	Synthesis of tryptophan and auxin.	Alloway (2004); Brennan (2005)
	Reduction of the adverse effects of short periods of heat and salt stresses.	Disante et al. (2010); Tavallali et al. (2010)
Co	Inhibition of ethylene production.	Lau and Yang (1976)
	Role in salt tolerance.	Ibrahim et al. (1989)
Mn	Activation of enzymes such as decarboxylating malate dehydrogenase, isocitrate dehydrogenase, and nitrate reductase.	Mukhopadhay and Sharma (1991)
Mo	Regulatory component in the maintenance of nitrogen fixation in legumes.	Kaiser et al. (2005); Soetan et al. (2010)
	Integral part of molybdenum cofactor (Moco), which binds to molybdenum-requiring enzymes.	Bittner et al. (2001); Mendel and Haensch (2002); Kaiser et al. (2005)
Ni	Cofactor of enzymes involved in DNA biosynthesis and amino acid metabolism.	Arinola et al. (2008)
	Component of the enzyme urease.	Aydinalp and Marinova (2009)
Hg	No beneficial effect reported.	

Source: Srivastava, S. and Bhargava, A., Genetic diversity and heavy metal stress in plants, in Ahuja M.R. and Jain S.M. (Eds.), *Sustainable Development and Biodiversity*, Springer International Publishing, Cham, Switzerland, pp. 241–270, 2015.

Despite their importance, heavy metals are considered major polluters of soil ecosystem due to their toxic effect (chronic, subchronic, or acute) on plants grown on such soils. The roots absorb heavy metals, which on entry into the plant system induce changes at morphological, physiological, and molecular levels. These changes or, more specifically, toxicity varies according to the plant species, type of heavy metal, its concentration, chemical structure, and soil factors. In human beings, heavy metals may induce several harmful effects, ranging from irritation to being toxic, mutagenic, teratogenic, or carcinogenic.

9.3 HYPERACCUMULATORS

Heavy metals have widespread occurrence and are considered soil pollutants due to their acute and chronic toxic effects on plants. It had long been observed that the uptake of metals by plants was selective, with some being preferentially acquired over others. Plants growing on metalliferous soils have been generally grouped into three categories:

1. *Excluders*: Plants that limit the levels of heavy metals by maintaining them up to a critical value across a wide range of soil concentrations.
2. *Accumulators*: Accumulators do not prevent metals from entering the roots and concentrate them in the aerial portions from low to high soil concentrations.
3. *Indicators*: Plants in which the internal concentration reflects the metal levels in the soil, but continued uptake of metals leads to mortality of the plant.

Although the ability of some plant species to accumulate heavy metals was known for quite some time, the term *hyperaccumulators* was introduced by Brooks et al. (1977) to refer to those plant species that can uptake metals, transport them to aerial parts of the plant, and accumulate up to 100-fold greater amount of the metal as compared with the nonaccumulator plants. Metal hyperaccumulators have the unique quality of accumulating heavy metals in their above-ground tissues, without concomitant development of any toxicity symptoms. Metal hyperaccumulators have been reported to occur in more than 450 species of vascular plants spread across 45 angiosperm families (Table 9.2), including members of the Brassicaceae, Asteraceae, Caryophyllaceae, Cyperaceae, Cunouniaceae, Fabaceae, Flacourtiaceae, Lamiaceae, Poaceae, Violaceae, and Euphorbiaceae, with the highest number found in Brassicaceae (approximately 11 genera and 87 species).

TABLE 9.2
Important Plant Species That Are Metal Hyperaccumulators

Metal	Number of Hyperaccumulator Species	Plant Species That Accumulate Specific Metals	Family	Reference
Cd	2	*Thlaspi caerulescens*	Brassicaceae	Basic et al. (2006)
		Arabis gemmifera	Brassicaceae	Kubota and Takinaka (2003)
		Arabidopsis halleri	Brassicaceae	Dahmani-Muller et al. (2000); Bert et al. (2002)
		Bidens pilosa	Asteraceae	Sun et al. (2009)
Ni	320	*Berkheya coddii*	Asteraceae	Robinson et al. (1997a); Moradi et al. (2010)
		Alyssum bertolonii	Brassicaceae	Robinson et al. (1997b)
		Sebertia acuminata	Sapotaceae	Jaffre et al. (1976)
		Phidiasia lindavii	Acanthaceae	Reeves et al. (1999)
Mn	9	*Austromyrtus bidwillii*	Myrtaceae	Bidwell et al. (2002)
		Phytolacca acinosa	Phytolaccaceae	Xue et al. (2004)
		Virotia neurophylla	Proteaceae	Fernando et al. (2006)
		Gossia bidwillii	Myrtaceae	Fernando et al. (2007)
		Maytenus founieri	Celastraceae	Fernando et al. (2008)
Cr	–	*Salsola kali*	Amaranthaceae	Gardea-Torresdey et al. (2005)
		Leersia hexandra	Poaceae	Zhang et al. (2007)
Pb	14	*Sesbania drummondii*	Fabaceae	Sahi et al. (2002); Sharma et al. (2004)
		Hemidesmus indicus	Apocynaceae	Chandra Shekhar et al. (2005)
Cu	34	*Ipomea alpina*	Convolvulaceae	Cunningham and Ow (1996)
		Crassula helmsii	Crassulaceae	Küpper et al. (2009)
Co	34	*Haumaniastrum robertii*	Lamiaceae	Brooks (1998)
		Crotalaria cobalticola	Fabaceae	Oven et al. (2002)
Zn	18	*Thlaspi caerulescens*	Brassicaceae	Brown et al. (1995)
		Arabis gemmifera	Brassicaceae	Kubota and Takinaka (2003)
		Sedum alfredii	Crassulaceae	Sun et al. (2005)
		Arabidopsis halleri	Brassicaceae	Zhao et al. (2000)
Se	20	*Astragalus bisulcatus*	Fabaceae	Beath et al. (1937)
		Brassica juncea	Brassicaceae	Orser et al. (1999)
		Stanleya pinnata	Brassicaceae	Parker et al. (2003)
Tl	–	*Iberis intermedia*	Brassicaceae	Leblanc et al. (1999)
		Brassica oleracea	Brassicaceae	Al-Najar et al. (2005)

Source: Bhargava, A. et al., *J. Environ. Manage.*, 105, 103–120, 2012.

9.4 PHYTOREMEDIATION AND THE ORIGIN OF PHYTOMINING

The term phytoremediation refers to the remediation technology that utilizes plant systems and soil microbes to efficiently remove inorganic and organic pollutants or render them harmless. This technique is generally applied to reduce the amount of pollutants present in water or soil. The word phytoremediation has originated from two words: the Greek word *phyto* meaning plant and the Latin suffix *remedium* meaning restoration or curing. Phytoremediation is an efficient, ecofriendly, solar-driven technology that is cost-effective and has good public acceptance. Phytoextraction is a subbranch of phytoremediation that refers to the removal of contaminants by utilization of hyperaccumulator plants, which concentrate contaminants in their above-ground parts. This accumulation is facilitated by the transportation and concentration of the contaminants in the above-ground harvestable parts. Phytoextraction is an ideal technique for the remediation of large tracts of land contaminated with heavy metals present at shallow depths. Phytoextraction has been receiving increasing attention for decontaminating metal-polluted soils. Phytomining involves the commercial aspect of phytoextraction, wherein hyperaccumulator plants are harvested or mined for minerals by cultivation on metal-rich soils. In comparison with conventional mining, where metals are obtained by digging the earth and refinement, phytomining uses hyperaccumulator plants to accumulate metals. When the metals have been uptaken and concentrated in the plant's tissue, the plants are harvested and left for drying. Thereafter, the plants are ground and burnt to obtain the metals. The dried plant material is reduced to bio-ore and thereafter treated by roasting, sintering, or smelting, which allows the metal recovery by conventional metal-refining methods such as acid dissolution and/or electrowinning. Figure 9.1 presents an overview of the phytomining process.

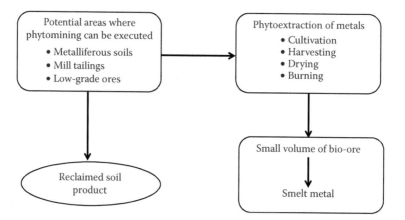

FIGURE 9.1 Extraction of metals by phytomining. (Reprinted from *Miner. Eng.*, 22, Sheoran, V. et al., Phytomining: A review, 1007–1019, Copyright 2009, with permission from Elsevier.)

9.5   HISTORY OF PHYTOMINING

The history of phytomining can be traced back to the early 1970s, when T. Jaffre noticed a tree *Sebertia acuminata* (Family: Sapotaceae) (locally known as *seve bleue*), which exuded a sap containing 26% nickel. However, it was Baker and Brooks (1989) who first propped the idea of growing hyperaccumulator plants for phytomining. Finally, in 1995, Nicks and Chambers, while working on the herb *Streptanthus polygaloides* (Family: Brassicaceae), a nickel hyperaccumulator, over nickel-rich *serpentine* soils at Chinese Camp in California, demonstrated the economic feasibility of growing a crop for nickel mining. It was found that *S. polygaloides* was capable of producing about 100 kg/ha of nickel, and incineration of the plant material yielded the bio-ore containing about 15% nickel and had the additional advantage of being sulfur-free. They concluded that the potential value of a crop of nickel was about the same as that of wheat, provided that some of the energy used in burning of the dry material could be utilized to yield about 15% of nickel metal. Two years later, Robinson et al. (1997b), while working in Tuscany, Italy, on the hyperaccumulator *Alyssum bertolonii* (Family: Brassicaceae) in phytomining of ultramafic soils for nickel, observed that moderate fertilization gave a threefold biomass increase, without concomitant loss of nickel. A nickel content of 0.8% in dry matter (11% in ash) gave a nickel yield of 72 kg/ha without the need of resowing of the further crop. In 1998, Chaney and coworkers came out with a patent on phytomining for nickel.

9.6   METHODOLOGY

The initial step in phytomining is the identification of land with the target metal; that is, the soil needs to have a minimum concentration of metal, so that the operation is economically viable. Once such a land is available, the hyperaccumulators are farmed and fertilizers are applied to increase the biomass of the hyperaccumulator crops. For example, in *Berkheya coddii*, a dry biomass of 22 t/ha could be achieved after moderate fertilization. Sometimes, metal uptake is also influenced by fertilizer application. Addition of fertilizers decreases the pH of the soil, which in turn enhances the bioavailability of metals, along with the growth of plants. Plants have shown enhanced uptake of nickel with increasing nitrogen addition on application of nitrogenous and phosphorus fertilizers. Besides fertilizer application, soil pH is known to influence metal uptake by plants. In recent years, much work has been done with respect to the use of metal-chelating compounds called phytochelators, which help transport heavy metals into plant tissues by increasing the bioavailability of metals that are tightly bound to the soil. Chelating agents such as ethylenediaminetetraacetic acid (EDTA), nitrilotriacetic acid (NTA), hydroxyethyliminodiacetic acid (HEIDA), citric acid (CA), diethylenetriaminepentaacetic acid (DTPA), ethylenediaminedisuccinic acid (EDDS), and thiocyanates form metal organic complexes that are water-soluble, which bring metals into solution through desorption of sorbed species. Agronomic practices are also known to influence phytomining. An interesting study (Bani et al. 2015) evaluated the effect of planting density (transplanted seedlings) on a nickel phytomining cropping system with native *Alyssum murale* (a perennial dicot herb) on two representative vertisols. The researchers concluded that a density of 4 plants/m² was suitable for phytoextraction of nickel by *A. murale*.

9.7 PHYTOMINING OF SPECIFIC METALS

9.7.1 NICKEL

Nickel (Ni) is a rare metal, whose main ores are nickel sulfides and nickel laterites, the tropical soils on top of olivine- or serpentine-rich rocks. About 400 nickel hyperaccumulators are distributed across 40 plant families. Most Ni hyperaccumulator plants accumulate 0.1%–0.5% Ni in their biomass, but for phytomining, *hypernickelophores* (>1% Ni) are used (Table 9.3).

Alyssum murale and *Allium corsicum* are reported to accumulate higher than 20,000 mg/kg nickel in shoot dry weight, with no evidence of phytotoxicity when grown on serpentine soils with minimal addition of fertilizers. In nickel phytomining operations, a selected hyperaccumulator plant species having high biomass yield and high Ni concentrations (>1%) in the above-ground biomass is grown on Ni-rich (ultramafic) soils, followed by harvesting and incineration of the biomass to produce a *bio-ore* from which Ni salts or Ni metal may be recovered. Good plant and soil management practices, based on experiences from laboratory and field tests, are required to maximize the yield. Application of conventional nitrogen, phosphorus, and potassium (NPK) fertilizer, addition of organic matter, and

TABLE 9.3

Plants with Scope for Nickel Phytomining

Plant	Family	Distribution	Shoot Ni (%)	Reference
Alyssum spp.	Brassicaceae	South and Southeast Europe, Turkey, Armenia, Iraq, and Syria	1–2.5	Brooks (1998)
Berkheya coddii	Asteraceae	South Africa and Zimbabwe	1.1	Morrey et al. (1989)
Bornmuellera spp.	Brassicaceae	Greece, Albania, and Turkey	1–3	Reeves et al. (1983)
Buxus spp.	Buxaceae	Cuba	1–2.5	Reeves et al. (1996)
Leucocroton spp.	Euphorbiaceae	Cuba	1–2.7	Reeves et al. (1996)
Leptoplax spp.	Brassicaceae	Greece	1–3.5	Reeves et al. (1980)
Pearsonia metallifera	Fabaceae	Zimbabwe	1.4	Wild (1974)
Phyllanthus spp.	Phyllanthaceae	Southeast Asia and Central America	2–6	Baker et al. (1992); van der Ent et al. (2015a)
Rinorea bengalensis	Violaceae	Southeast Asia	1–2.7	Brooks and Wither (1977) Jopony and Tongkul (2011)

substitution of plant growth regulators significantly increase the biomass yield of *metal crops*, without causing dilution in shoot Ni yields. A recent study published in 2016 has shown that highest biomass production was generally found after addition of 2.5% or 5% compost. Addition of finely ground elemental sulfur lowers the soil pH, thereby increasing the NH_4-acetate-extractable Ni fraction. Calcium supply in the form of $CaCO_3$ increases Ni uptake due to the result of combined effect of Ca addition and soil pH. Since Ni phytomining has similar costs of production as some food crops, Ni phytomining has become a viable business opportunity for *metal farmers*, especially in some developing countries, and could emerge as a profitable agricultural industry.

9.7.2 GOLD

Early reports from the twentieth century discussed about the accumulation of gold by plants, especially the conifers that can concentrate this precious metal in the order of parts per billion in their tissues. Gold, in its natural state, is highly insoluble, which reduces its bioavailability and limits the potential for phytoextraction, since bioavailability is one of the critical factors affecting metal uptake by plants. Thus, plants normally do not concentrate gold, and the metal has to be made soluble before uptake can occur. No plant has been reported so far that has a physiological mechanism to promote gold solubility at the soil-root interface. Table 9.4 provides details of some important studies undertaken with reference to gold phytoextraction. To make gold extraction viable, the harvesting of crop should produce a dry biomass of 10 t from 1 ha of land and there should be a gold concentration of 100 mg/kg in the dry biomass. This would yield 1 kg of gold per hectare.

TABLE 9.4
Studies on Gold Phytoextraction

Plant	Type of Trial	Reference
Brassica juncea	Laboratory	Anderson et al. (1998a, 1998b)
Medicago sativa	Laboratory	Gardea-Torresdey et al. (2002)
Raphanus sativus	Laboratory	Msuya et al. (2000)
Allium cepa	Laboratory	Msuya et al. (2000)
Beta vulgaris	Laboratory	Msuya et al. (2000)
Daucus carota	Laboratory	Msuya et al. (2000)
Chilopsis linearis	Laboratory	Gardea-Torresdey et al. (2002)
Trifolium repens	Laboratory	Piccinin et al. (2007)
Brassica campestris	Greenhouse	Wilson-Corral (2008)
Sorghum halepense	Greenhouse	Rodriguez-Lopez et al. (2009a)
Kalanchoe serrata	Greenhouse	Rodriguez-Lopez et al. (2009b)
Brassica juncea	Field	Anderson et al. (2005)
		Wilson-Corral (2008)

Gold phytomining process involves the following five steps or stages (Anderson et al. 2003):

Stage one: Identification of a gold mining site or auriferous area having soil rich in gold

Stage two: Planting a fast-growing plant species having large biomass, which is tolerant of the dry, acidic, and/or saline conditions

Stage three: Treating the soil with a chosen chemical, as the plants approach maximum biomass near maturity

Stage four: Harvesting the plants as they start showing signs of poor health due to metal shock and chemical toxicity

Stage five: Recovering the gold from the biomass

Cyanide, a highly poisonous chemical that has a requirement for iron-based *heme* compounds for oxygen transport, is the primary lixiviant used to dissolve gold by the mining industry. Cyanide is nontoxic to plants that have no requirement for iron-based oxygen transport compounds. Although there are environmental concerns about adding chemical agents to the soil, cyanide gas is emitted naturally by some plants such as soybean, which can be planted near hyperaccumulators to dissolve the gold in the soil. This is a more environmentally friendly way to phytomine gold.

9.7.3 Rhenium

Rhenium (Re) is a rare metal with ultra-high melting point (3180°C), extremely high density of 21.04 g/cm³, and extremely high recrystallization temperature (2800°C), which gives it excellent creep resistance and high ductility over a wide temperature range, due to which it is used in high-temperature superalloy turbine blades in jet engines, land-based gas-powered turbines, and aerospace industry. The conventional methods for obtaining rhenium from main carrier minerals contaminate the environment with acid emissions and chemical solutions, and the scattered nature of the metal makes it very expensive to collect it from the environment of ore dressing and work processing regions by classical technology. Rhenium is preferably accumulated (>98%) in green above-ground parts of all kinds of plants. Bozhkov and Tzvetkova (2009) developed a low-cost technological scheme for phytomining of rhenium from ore-dressing soils, with a number of advantages over classical methods for rhenium recovery. A most recent report (Novo et al. 2015) assessed the feasibility of Re phytomining by using Indian mustard (*Brassica juncea*; Family: Brassicaceae). The results showed high concentrations of Re in plants, ranging from 1553 to 22,617 mg/kg at 45 days and from 1348 to 23,396 mg/kg at 75 days. A profit of USD 3906/ha harvest was demonstrated from the recovered Re, which demonstrated for the first time the scientific and economic viability of Re phytomining.

9.7.4 Thallium

Thallium, a metal serendipitously discovered by Sir William Crookes, is of immense industrial importance and is used in the manufacture of alloys, semiconductor materials, electronic components, optical lenses, gamma radiation detection equipment, low-temperature thermometers, imitation jewelry, and green fireworks. Thallium is a soft, colorless, odorless, and tasteless pliable metal with an atomic mass of 204.38 and a melting point of 303.5°C. Thallium is a highly toxic metal, with a crustal abundance of about 0.7 mg/kg (ppm). A large number of food crops are known to accumulate thallium, notably the members of family Brassicaceae, especially *Iberis intermedia, Biscutella laevigate*, and *Brassica napus*. Studies have shown that it could be possible to produce 8 kg of thallium (worth USD 2400) from 1 ha of land containing 10 mg/kg thallium planted with *I. intermedia*. Using this scenario, it seems possible to produce a crop worth about USD 180 from 1 ha of *Iberis*, which would yield 600 g/ha of thallium. To be economical, a crop with a biomass of 10 t/ha would have to contain at least 170 mg/kg thallium in dry matter. Although presently less economical, the results show bright prospects for thallium phytomining during the high-demand period (market price USD 300/kg) if sufficiently large tracts of thallium-contaminated lands are at disposal, in order to obtain the advantage of large-scale production.

9.8   ECONOMIC FEASIBILITY

The economic feasibility of phytomining ultimately depends on the market price of the metal, the annual yield per unit area, metal content of the plant, and the availability of surface areas enriched in the target element. Researchers from the University of Sydney have evaluated the economic viability of gold phytomining in Australia. For gold, a profit of AUS$ 26,000/ha/harvest has been envisaged using thiocyanate-induced accumulation with a crop of Indian mustard (*B. juncea* L.), an annual perennial herb, coupled with energy generation from the harvested biomass.

Presently, large-scale harvesting of metal-accumulated plants is more costly as compared with the extraction of metals from mines by using conventional mining techniques. However, phytomining could become a cost-effective alternative to conventional mining if the scale of production is large enough. This scenario is likely to change in the future, as the yields from mines deplete, which would lead to an increase in metal prices. The increasing demand for metals by the industry and their shortage would help in offsetting the costs of initiating large-scale phytomining farm production. According to the present status, for phytomining to be ever realized as a commercial enterprise, cobalt and nickel will be the most profitable entities, which lie in the price range of USD 6090–USD 48,000 per ton and which are known to have many plant hyperaccumulators.

9.9   ADVANTAGES OF PHYTOMINING

Phytomining is generally considered a green technology as compared with the conventional mining practices. It can be considered a viable alternative to opencast mining of low-grade ores. Phytominers can recollect metal pollutants from metal-contaminated soils, thereby restoring the soil to its original health and vigor. Bio-ores are generally sulfur-free, and their smelting requires less energy than sulfidic ores, which deteriorate the environment by causing acid rains. The bio-ores also have greater metal content than the conventional ores and therefore require less space for storage. However, the growing mass amount of plants also takes a toll on the land used for cultivation. The farming practices suited for phytomining deplete the soil, and overgrowing of biocrops can permanently alter the ecology of a particular area. Such are the economic potentials of phytomining for farming communities that van der Ent et al. (2015b) have proposed a new term *agromining* (a variant of phytomining), which could provide local communities with an alternative type of agriculture on degraded lands by cultivation of crops for metals instead of farming of food crops.

9.10   LIMITATIONS

Phytomining allows metals to be extracted from soils that are considered uneconomic by conventional mining methods and therefore should not be compared to the production rates of conventional mining. For example, nickel can be phytomined from soils containing as little as 0.5% nickel. Thus, phytomining does not compete but rather supplements conventional mining. Another limitation of phytomining is the inability of plant roots to penetrate deep into the soil. Owing to this, the hyperaccumulators are able to extract metals only on the surface of the soil. Thus, phytomining is restricted in its use with respect to deep-seated deposits. Another drawback is that in contrast to conventional mining, the success of phytomining is subject to the forces of nature such as the weather, altitude, and soil quality. One erratic growing season is detrimental for phytomining, since an entire crop of metal-producing plants can be destroyed, and if weather patterns are altered by global climate change, there would be greater risk in establishing a long-lasting phytomining industry.

The agronomic limitations such as supplementing soils with fertilizers, compost, and irrigation need special mention. For example, ultramafic soils, where Ni phytomining would occur, have poor soil structure and are usually nutrient-poor. These soils have large requirements for fertilizers and irrigation because of their inability to retain nutrients and water. The use of soil amendments such as compost improves the water- and nutrient-holding capacities but may be regressive in lowering the availability of the target metal, which in turn is likely to reduce profits. In addition, seed supply of hyperaccumulator plants is a constraint, since no large-scale suppliers of seeds are available that would be used in phytomining.

Another controversy relates to the long-term effects of phytomining. The chief issue pertains to the allocation of agricultural land for the phytomining industry when humans have only limited amount of good farming land available. The effect of having metal-enhanced plants entering the food chain over time also needs to be evaluated, along with the assessment of the fact of possible metal runoff from the plants entering the local water supply.

9.11   FUTURE PERSPECTIVES

Phytoextraction and phytomining have faced innumerable challenges in finding widespread application for decontaminating soil in the last two decades. However, increasing research for improvement of metal uptake is the need of the hour, since phytomining has the potential for widespread applications in various sectors that will benefit mankind. Phytomining can be combined with other technologies such as mine rehabilitation, in which case the revenue generated from phytomining can offset the cost of remediation. In this scenario, plants should be used to recover metals from waste rock or tailings, which have metal concentrations too low for economic harnessing by conventional methods. One advantage of following this procedure is that the infrastructure to recover the metals would be already in place. Another advantage relates to the phytomining operation, creating opportunities for imparting training to rural communities in modern agriculture, which would be essential in ensuring that targets are met with respect to phytomining, and getting farmers well acquainted and skilled in the cultivation of specific crop species on nonmineralized land.

BIBLIOGRAPHY

Alloway B.J. 2004. *Zinc in Soil and Crop Nutrition*. International Zinc Association, Brussels, Belgium.

Al-Najar H., Kaschl A., Schulz R. and Romheld V. 2005. Effects of thallium fractions in the soil and pollution origin in thallium uptake by hyperaccumulator plants: A key factor for assessment of phytoextraction. *Int. J. Phytoremediation* 7: 55–67.

Álvarez-López V., Prieto-Fernández Á., Cabello-Conejo M.I. and Kidd P.S. 2016. Organic amendments for improving biomass production and metal yield of Ni-hyperaccumulating plants. *Sci. Total Environ.* 548–549: 370–379.

Anderson C., Moreno F. and Meech J. 2005. A field demonstration of gold phytoextraction technology. *Miner. Eng.* 18: 385–392.

Anderson C.W.N., Brooks R., Stewart R., Simcock R. and Robinson B. 1999a. The phytoremediation and phytomining of heavy metals. *Pacrim 99*, Ball, Indonesia, pp. 127–135.

Anderson C.W.N., Brooks R.R., Chiarucci A., LaCoste C.J., Leblanc M., Robinson B.H., Simcock R. and Stewart R.B. 1999b. Phytomining for nickel, thallium and gold. *J. Geochem. Explor.* 67: 407–415.

Anderson C.W.N., Brooks R.R., Stewart R.B. and Simcock R. 1998a. Harvesting a crop of gold in plants. *Nature* 395: 553–554.

Anderson C.W.N., Brooks R.R., Stewart R.B. and Simcock R. 1998b. Gold uptake by plants. *Gold Bull.* 32: 48–51.

Anderson C.W.N., Moreno F. and Meech J. 2005. A file demonstration of gold phytoextraction technology. *Miner. Eng.* 18: 385–392.

Anderson C., Stewart R., Moreno F., Wreesman C., Gardea-Torresday J., Robinson B. et al. 2003. Gold phytomining. Novel developments in a plant-based mining system. *Proceedings of the Gold 2003 Conference: New Industrial Applications of Gold*. World Gold Council and Canadian Institute of Mining, Metallurgy and Petroleum, Vancouver, Canada.

Arinola O.G., Nwozo S.O., Ajiboye J.A. and Oniye A.H. 2008. Evaluation of trace elements and total antioxidant status in Nigerian cassava processors. *Pakistan J. Nutr.* 7: 770–772.

Aydinalp C. and Marinova S. 2009. The effect of heavy metals on seed germination and plant growth on alfalfa plant (*Medicago sativa*). *Bulgarian J. Agr. Sci.* 15: 347–350.

Baker A.J.M. 1981. Accumulators and excluders—strategies in the response of plants to heavy metals. *J. Plant Nutr.* 3: 643–654.

Baker A.J.M. and Brooks R.R. 1989. Terrestrial higher plants which hyperaccumulate metallic elements. A review of their distribution, ecology and phytochemistry. *Biorecovery* 1: 81–126.

Baker A.J.M., Proctor J., Van Balgooy M. and Reeves R. 1992. Hyperaccumulation of nickel by the flora of the ultramafics of Palawan, Republic of the Philippines. In: Baker A.J.M., Proctor J. and Reeves R.D. (Eds.). *The Vegetation of Ultramafic (Serpentine) Soils: Proceedings of the First International Conference on Serpentine Ecology*. Intercept, Andover, UK.

Bani A., Echevarria G., Zhang X., Benizri E., Laubie B., Morel J.L. and Simonnot M.O. 2015. The effect of plant density in nickel-phytomining field experiments with *Alyssum murale* in Albania. *Aust. J. Bot.* 63: 72–77.

Barzanti R., Colzi I., Arnetoli M., Gallo A., Pignattelli S., Gabbrielli R. and Gonnelli C. 2011. Cadmium phytoextraction potential of different *Alyssum* species. *J. Hazard. Mater.* 196: 66–72.

Basic N., Keller C., Fontanillas P., Vittoz P., Besnard G. and Galland N. 2006. Cadmium hyperaccumulation and reproductive traits in natural *Thlaspi caerulescens* populations. *Plant Biol.* 8: 64–72.

Beath O.A., Eppsom H.F. and Gilbert C.S. 1937. Selenium distribution in and seasonal variation of type vegetation occurring on seleniferous soils. *J. Pharm. Assoc.* 26: 394–405.

Becerra-Castro C., Monterroso C., Garcia-Leston M., Prieto-Fernandez A., Acea M.J. and Kidd P.S. 2009. Rhizosphere microbial densities and trace metal tolerance of the nickel hyperaccumulator *Alyssum serpyllifolium* subsp. *lusitanicum*. *Int. J. Phytoremediation* 11: 525–541.

Bech J., Duran P., Roca N., Poma W., Sánchez I., Barceló J., Boluda R., Roca-Pérez L. and Poschenrieder C. 2012. Shoot accumulation of several trace elements in native plant species from contaminated soils in the Peruvian Andes. *J. Geochem. Explor.* 113: 106–111.

Bert V., Bonnin I., Saumitou-Laprade P., de Laguerie P. and Petit D. 2002. Do *Arabidopsis halleri* from nonmetallicolous populations accumulate zinc and cadmium more effectively than those from metallicolous populations? *New Phytol.* 155: 47–57.

Bhargava A. and Srivastava S. 2014. Transgenic approaches for phytoextraction of heavy metals. In: Ahmad P., Wani M.R., Azooz M.M. and Tran L.P. (Eds). *Improvement of Crops in the Era of Climatic Changes*. Springer, New York. pp. 57–80.

Bhargava A., Carmona F.F., Bhargava M. and Srivastava S. 2012a. Approaches for enhanced phytoextraction of heavy metals. *J. Environ. Manage.* 105: 103–120.

Bhargava A., Gupta V.K., Singh A.K. and Gaur R. 2012b. Microbes for heavy metal remediation. In: Gaur R., Mehrotra S. and Pandey R.R. (Eds.). *Microbial Applications*. IK International Publishing, New Delhi, India, pp 167–177.

Bhargava A., Shukla S., Srivastava J., Singh N. and Ohri D. 2008. *Chenopodium*: A prospective plant for phytoextraction. *Acta Physiol. Plant.* 30: 111–120.

Bidwell S.D., Woodrow I.E., Batianoff G.N. and Sommer-Knudsen J. 2002. Hyperaccumulation of manganese in the rainforest tree *Austromyrtus bidwillii* (Myrtaceae) from Queensland, Australia. *Funct. Plant Biol.* 29: 899–905.

Bittner F., Oreb M. and Mendel R.R. 2001. ABA3 is a molybdenum cofactor sulfurase required for activation of aldehyde oxidase and xanthine dehydrogenase in *Arabidopsis thaliana*. *J. Biol. Chem.* 276: 40381–40384.

Bozhkov O. and Tzvetkova C. 2009. Advantages of Rhenium Phytomining by Lucerne and Clover from Ore Dressing Soils. In: *Proceedings 7th WSEAS International Conference on Environment, Ecosystems and Development (EED '09)*. WSEAS Press, Puerto De La Cruz, Spain, pp. 127–131.

Brennan R.F. 2005. Zinc application and its availability to plants. PhD dissertation, School of Environmental Science, Division of Science and Engineering, Murdoch University.

Brooks R.R. 1987. *Serpentine and its Vegetation: A Multidisciplinary Approach*. Discorides Press, Portland, OR, p. 454.

Brooks R.R. 1998. *Plants that Hyperaccumulate Heavy Metals*. CAB International, Wallingford, UK.

Brooks R.R. and Robinson B.H. 1998. The potential use of hyperaccumulators and other plants for phytomining. In: Brooks R.R. (Ed.). *Plants that Hyperaccumulate Heavy Metals: Their Role in Archaeology, Microbiology, Mineral Exploration, Phytomining and Phytoremediation*. CAB International, Wallingford, UK, pp. 327–356.

Brooks R.R. and Wither E.D. 1977. Nickel accumulation by *Rinorea bengalensis* (Wall.) O.K. *J. Geochem. Explor.* 7: 295–300.

Brooks R.R., Anderson C., Stewart R. and Robinson B. 1999. Phytomining: Growing a crop of a metal. *Biologist* 46: 201–205.

Brooks R.R., Chambers M.F., Nicks L.J. and Robinson B.H. 1998. Phytomining. *Trends Plant Sci.* 3: 359–362.

Brooks R.R., Lee J., Reeves R.D. and Jaffre T. 1977. Detection of nickeliferous rocks by analysis of herbarium specimens of indicator plants. *J. Geochem. Explor.* 7: 49–57.

Brown S.L., Chaney R.F., Angle J.S. and Baker A.J.M. 1995. Zn and Cd uptake by hyper-accumulator *Thlaspi caerulescens* and metal tolerant *Silene vulgaris*, grown on sludge- amended soils. *Environ. Sci. Tech.* 29: 1581–1585.

Chandra S.K., Kamala C.T., Chary N.S., Balaram V. and Garcia G. 2005. Potential of *Hemidesmus indicus* for phytoextraction of lead from industrially contaminated soils. *Chemosphere* 58: 507–514.

Chaney R.F., Angle J.S., Baker A.J.M., Reeves R.D., Roseberg R.J., Simmons R.W. and Broadhurst C.L. 2008. Phytoextraction and phytomining of Ni and Cd from contami-nated or mineralized soils. In: Bert V. (Ed.). *COST Action 859, Phytotechnologies in Practice: Biomass Production, Agricultural Methods, Legacy, Legal and Economic Aspects.* COST, Verneuil-en-Halatte, France, pp. 15–16.

Chaney R.L., Angle J.S., Baker A.J.M. and Li J.M. 1998. Method for phytomining of nickel, cobalt, and other metal from soil. US Patent 5, 711: 784.

Chaney R.L., Angle J.S., Broadhurst C.L., Peters C.A., Tappero R.V. and Sparks D.L. 2007. Improved understanding of hyperaccumulation yields commercial phytoextraction and phytomining technologies. *J. Environ. Qual.* 36: 1429–1443.

Chaney R.L., Li Y.M., Brown S.L., Homer F.A., Malik M., Angle J.S., Baker A.J.M., Reeves R.D. and Chin M. 2000. Improving metal hyperaccumulator wild plants to develop commercial phytoextraction systems: Approaches and progress. In: Terry N. and Banuelos G. (Eds.). *Phytoremediation of Contaminated Soil and Water.* Lewis Publishers, Boca Raton, FL, pp. 129–158.

Chaney R.L., Malik M., Li Y.-M., Brown S.L., Brewer E.P., Angle J.S. and Baker A.J.M. 1997. Phytoremediation of soil metals. *Curr. Opin. Biotech.* 8: 279–284.

Cheng S. 2003. Heavy metals in plants and phytoremediation. *Environ. Sci. Pollut. Res.* 10: 335–340.

Cunningham S.D. and Ow D.W. 1996. Promises and prospects of phytoremediation. *Plant Physiol.* 110: 715–719.

Dahmani-Muller H., van Oort F., Gelie B. and Balabane M. 2000. Strategies of heavy metal uptake by three plant species growing near a metal smelter. *Environ. Pollut.* 109: 231–238.

Demirevska-kepova K., Simova-Stoilova L., Stoyanova Z., Holzer R. and Feller U. 2004. Biochemical changes in barely plants after excessive supply of copper and manganese. *Environ. Exp. Bot.* 52: 253–266.

Disante K.B., Fuentes D. and Cortina J. 2010. Response to drought of Zn-stressed *Quercus suber* L. Seedlings. *Env. Exp. Bot.* 70: 96–103.

Du R.J., He E.K., Tang Y.T., Hu P.J., Ying R.R., Morel J.L. and Qiu R.L. 2011. How phytohor-mone IAA and chelator EDTA affect lead uptake by Zn/Cd hyperaccumulator *Picris divaricata. Int. J. Phytoremediation* 13: 1024–1036.

Dunn C.E. 1995. Biogeochemical prospecting for metals. In: Brooks R.R., Dunn C.E. and Hall G.E.M. (Eds.). *Biological Systems in Mineral Exploration and Processing.* Ellis Horwood, Hemel Hemptead, UK, pp. 371–426.

Fernando D.R., Bakkaus E.J., Perrier N., Baker A.J.M., Woodrow I.E., Batianoff G.N. and Collins R.N. 2006. Manganese accumulation in the leaf mesophyll of four tree species: A PIXE/EDAX localization study. *New Phytol.* 171: 751–758.

Fernando D.R., Woodrow I.E., Bakkaus E.J., Collins R.N., Baker A.J.M. and Batainoff G.N. 2007. Variability of Mn hyperaccumulation in the Australian rainforest tree *Gossia bidwillii* (Myrtaceae). *Plant and Soil* 293: 145–152.

Fernando D.R., Woodrow I.E., Jaffre T., Dumontet V., Marshall A.T. and Baker A.J.M. 2008. Foliar manganese accumulation by *Maytenus founieri* (Celastraceae) in its native New Caledonian habitats: Populational variation and localization by X-ray microanalysis. *New Phytol.* 177: 178–185.

Freeman J.L., Tamaoki M. and Stushnoff C. 2010. Molecular mechanisms of selenium tolerance and hyperaccumulation in *Stanleya pinnata*. *Plant Physiol.* 153: 1630–1652.

Galeas M.L., Zhang L.H., Freeman J.L., Wegner M. and Pilon-Smits E.A.H. 2007. Seasonal fluctuations of selenium and sulfur accumulation in selenium hyperaccumulators and related nonaccumulators. *New Phytol.* 173: 517–525.

Gardea-Torresday J.L., De la Rosa G., Peralta-Videa J.R., Montes M., Cruz-Jimenez G. and Cano-Aguilera I. 2005a. Differential uptake and transport of trivalent and hexavalent chromium by tumble wed (*Salsola kali*). *Arch. Environ. Contam. Toxicol.* 48: 225–232.

Gardea-Torresdey J.L., Parsons J.G., Gomez E., Peralta-Videa J., Troiani H.E., Santiago P. and Jose-Yacaman M. 2002. Formation and growth of Au nanoparticles inside live alfalfa plants. *Nano Lett.* 2: 397–401.

Gardea-Torresdey J.L., Rodriguez E., Parsons J.G., Peralta-Videa J.R., Meitzner G. and Cruz-Jimenez G. 2005b. Use of ICP and XAS to determine the enhancement of gold phytoextraction by *Chilopsis linearis* using thiocyanate as a complexing agent. *Anal. Bioanal. Chem.* 382: 347–352.

Harris A.T., Naidoo K., Nokes J., Walker T. and Orton F. 2009. Indicative assessment of the feasibility of Ni and Au phytomining in Australia. *J. Cleaner Prod.* 17: 194–200.

Hladun K.R., Parker D.R. and Trumble J.T. 2011. Selenium accumulation in the floral tissues of two Brassicaceae species and its impact on floral traits and plant performance. *Environ. Exper. Bot.* 74: 90–97.

Ibrahim A., El-Abd S. and El-Beltagy A.S. 1989. A possible role of cobalt in salt tolerance of plant. *Egypt J. Soil Sci.* 359–370.

Jaffre T., Brooks R.R., Lee J. and Reeves R.D. 1976. *Sebertia acuminata*: A hyperaccumulator of nickel from New Caledonia. *Science* 193: 579–580.

Jopony M. and Tongkul F. 2011. Heavy metal hyperaccumulating plants in Malaysia and its potential applications. In: Kuhn K. (Ed.). *New Perspectives in Sustainable Management in Different Woods.* Schriftenreihe der SRH Hochschule Heidelberg, Verlag Berlin GmbH. Logos Verlag Berlin GmbH, Berlin, Germany, pp. 129–142.

Kaiser B.N., Gridley K.L., Brady J.N., Phillips T. and Tyerman S.D. 2005. The role of molybdenum in agricultural plant production. *Ann. Bot.* 96: 745–754.

Kubota H. and Takenaka C. 2003. *Arabis gemmifera* is a hyperaccumulator of Cd and Zn. *Int. J. Phytoremediation* 5: 197–120.

Kupper H. and Kochian L.V. 2010. Transcriptional regulation of metal transport genes and mineral nutrition during acclimatization to cadmium and zinc in the Cd/Zn hyperaccumulator, *Thlaspi caerulescens* (Ganges population). *New Phytol.* 185: 114–129.

Küpper H., Mijovilovich A., Götz B., Kroneck P.M.H., Küpper F.C. and Meyer-Klaucke W. 2009. Complexation and toxicity of copper in higher plants (I): Characterisation of copper accumulation, speciation and toxicity in *Crassula helmsii* as a new copper hyperaccumulator. *Plant Physiol.* 151: 702–714.

Lamb A.E., Anderson C.W.N. and Haverkamp R.G. 2001. The extraction of gold from plants and its applications to phytomining. *Chem. New Zealand* 3: 1–33.

Lange B., van der Ent A., Baker A.J., Echevarria G., Mahy G., Malaisse F., Meerts P., Pourret O., Verbruggen N. and Faucon M.P. 2017. Copper and cobalt accumulation in plants: A critical assessment of the current state of knowledge. *New Phytol.* 213: 537–551.

Lasat M.M. 2002. Phytoextraction of toxic metals: A review of biological mechanisms. *J. Environ. Qual.* 31: 109–120.

Lau O. and Yang S.F. 1976. Inhibition of ethylene production by cobaltous ion. *Plant Physiol.* 58: 114–117.

Leblanc M., Petit D., Deram A., Robinson B. and Brooks R.R. 1999. The phytomining and environmental significance of hyperaccumulation of thallium by *Iberis intermedia* from southern France. *Econ. Geol.* 94: 109–113.

Li Y.M., Chaney R.L., Brewer E., Roseberg R., Angle J.S., Baker A., Reeves R. and Nelkin J. 2003. Development of a technology for commercial phytoextraction of nickel: Economic and technical considerations. *Plant and Soil* 249: 107–115.

Mendel R.R. and Haensch R. 2002. Molybdoenzymes and molybdenum cofactor in plants. *J. Exp. Bot.* 53: 1689–1698.

Miller G.W., Haung I.J., Welkie G.W. and Pushnik J.C. 1995. Function of iron in the plants with special emphasis on chloroplast and photosynthetic activity. In: Abadía J. (Ed.). *Iron Nutrition in Soils and Plants.* Kluwer, Dordrecht, the Netherlands, pp. 19–28.

Mongkhonsin B., Nakbanpote W., Nakai I., Hokura A. and Jearanaikoon N. 2011. Distribution and speciation of chromium accumulated in *Gynura pseudochina* (L.) DC. *Environ. Exp. Bot.* 74: 56–64.

Moradi A.B., Swoboda S., Robinson B., Prohaska T., Kaestner A., Oswald S.E., Wenzel W.W. and Schulin R. 2010. Mapping of nickel in root cross-sections of the hyperaccumulator plant *Berkheya coddii* using laser ablation ICP-MS. *Environ. Exp. Bot.* 69: 24–31.

Morrey D.R., Balkwill K. and Balkwill M.J. 1989. Studies on serpentine flora: Preliminary analyses of soils and vegetation associated with serpentinite rock formations in the Southeastern Transvaal. *S. Afr. J. Bot.* 55: 171–177.

Msuya F.A., Brooks R.R. and Anderson C. 2000. Chemically-induced uptake of gold by root crops: Its significance for phytomining. *Gold Bull.* 33: 134–137.

Mukhopadhay M.J. and Sharma A. 1991. Manganese in cell metabolism of higher plants. *Bot. Rev.* 57: 117–149.

Nicks L. and Chambers M.F. 1994. Nickel farm. *Discover*, September, p. 19.

Nicks L. and Chambers M.F. 1995. Farming for metals. *Mining Environ. Manage.* 3: 15–18.

Nkrumah P.N., Baker A.J.M., Chaney R.L., Erskine P.D., Echevarria G., Morel J.L. and van der Ent A. 2016. Current status and challenges in developing nickel phytomining: An agronomic perspective. *Plant Soil* 406: 55–69.

Novo L.A.B., Mahler C.F. and González L. 2015. Plants to harvest rhenium: Scientific and economic viability. *Environ. Chem. Lett.* 13: 439–445.

Orser C.S., Salt D.E., Pickering I.I., Epstein A. and Ensley B.D. 1999. *Brassica* plants to provide enhanced mineral nutrition: Selenium phytoenrichment and metabolic transformation. *J. Med. Food* 1: 253–261.

Oven M., Grill E., Golan-Goldhirsh A., Kutchan T.M. and Zenk M.H. 2002. Increase of free cysteine and citric acid in plant cells exposed to cobalt ions. *Phytochemistry* 60: 467–474.

Padmavathiamma P.K. and Li L.Y. 2007. Phytoremediation technology: Hyperaccumulation metals in plants. *Plant and Soil* 184: 105–126.

Parker D.R., Feist L.J., Varvel T.W., Thomason D.N. and Zhang Y.Q. 2003. Selenium phytoremediation potential of *Stanleya pinnata. Plant Soil* 249: 157–165.

Perrier N. 2004. Nickel speciation in *Sebertia acuminata*, a plant growing on a lateritic soil of New Caledonia. *Geoscience* 336: 567–577.

Piccinin R.C.R., Ebbs S.D., Reichman S.M., Kolev S.D., Woodrow I.E. and Baker A.J.M. 2007. A screen of some native Australian flora and exotic agricultural species for their potential application in cyanide-induced phytoextraction of gold. *Miner. Eng.* 20: 1327–1330.

Pollard A.J., Stewart H.L. and Roberson C.B. 2009. Manganese hyperaccumulation in *Phytolacca americana* L. from the southeastern united states. *Northeast. Nat.* 16: 155–162.

Reeves R.D. and Baker A.J.M. 2000. Metal accumulating plants. In: Raskin I. and Ensley B.D. (Eds.). *Phytoremediation of Toxic Metals: Using Plants to Clean Up the Environment.* Wiley, New York, pp. 193–229.

Reeves R.D., Adiguzel N. and Baker A.J.M. 2009. Nickel hyperaccumulation in *Bornmuellera kiyakii* and associated plants of the Brassicaceae from Kızıldaäÿ Derebucak (Konya), Turkey. *Turk. J. Bot.* 33: 33–40.

Reeves R.D., Baker A.J.M., Borhidi A. and Berazaín R. 1996. Nickel accumulating plants from the ancient serpentine soils of Cuba. *New Phytol.* 133: 217–224.

Reeves R.D., Baker A.J.M., Borhidi A. and Berazain R. 1999. Nickel hyperaccumulation in the serpentine flora of Cuba. *Ann. Bot.* 83: 29–38.

Reeves R.D., Brooks R.R. and Dudley T.R. 1983. Uptake of nickel by species of *Alyssum, Bornmuellera,* and other genera of Old World Tribus Alysseae. *Taxon* 32: 184–192

Reeves R.D., Brooks R.R. and Press J.R. 1980. Nickel accumulation by species of *Peltaria* Jacq. (Cruciferae). *Taxon* 29: 629–633.

Richards T. 2007. Guillain-Barre syndrome. Guillain-Barre syndrome fact sheet. National Institute of Neurological Disorders and Stroke, Bethesda, MD.

Robinson B.H., Anderson C.W.N. and Dickinson N.M. 2015. Phytoextraction: Where's the action? *J. Geochem. Explor.* 151: 34–40.

Robinson B.H., Brooks R.R., Howes A.W., Kirkman J.H. and Gregg P.E.H. 1997a. The potential of the high-biomass nickel hyperaccumulator *Berkheya coddii* for phytoremediation and phytomining. *J. Geochem. Explor.* 60: 115–126.

Robinson B.H., Chiarucci A., Brooks R.R., Petit D., Kirkman J.H., Gregg P.E.H. and De Dominicis V. 1997b. The nickel hyperaccumulator plant *Alyssum bertolonii* as a potential agent for phytoremediation and phytomining of nickel. *J. Geochem. Explor.* 59: 75–86.

Rodriguez-Lopez M., Wilson-Corral V., Anderson C. and Lopez-Perez J. 2009a. Chemically assisted gold phytoextraction in *Sorghum halepense.* In: *5th International Conference: Science, Technology and Applications of Gold,* Heidelberg, Germany.

Rodriguez-Lopez M., Wilson-Corral V., Lopez-Perez J., Arenas-Vargas M. and Anderson C. 2009b. Aplicacion Potencial de *Kalanchoe serrata* en Tecnologías de Mineria Ambientalmente Sostenibles. In: Barrios Durstewitz C.P. (Ed.). *Primer Congreso Internacional de Ciencias de la Ingeniería.* Universidad Autonoma de Sinaloa, Los Mochis, Mexico.

Sahi S.V., Bryant N.L., Sharma N.C. and Singh S.R. 2002. Characterization of a lead hyperaccumulator shrub, *Sesbania drummondii. Environ. Sci. Technol.* 36: 4676–4680.

Sharma N.C., Gardea-Torresday J.L., Parson J. and Sahi S.V. 2004. Chemical speciation of lead in *Sesbania drummondii. Environ. Toxicol. Chem.* 23: 2068–2073.

Sheoran V., Sheoran A.S. and Poonia P. 2009. Phytomining: A review. *Miner. Eng.* 22: 1007–1019.

Soetan K.O., Olaiya C.O. and Oyewole O.E. 2010. The importance of mineral elements for humans, domestic animals and plants: A review. *Afr. J. Food Sci.* 4: 200–222.

Spiller S.C., Castelfranco A.M. and Castelfranco P.A. 1982. Effects of iron and oxygen on chlorophyll biosynthesis: I. In vivo observations on iron and oxygen deficient plants. *Plant Physiol.* 69: 107–111.

Srivastava S. and Bhargava A. 2015. Genetic diversity and heavy metal stress in plants. In: Ahuja M.R. and Jain S.M. (Eds.). *Sustainable Development and Biodiversity.* Springer International Publishing, Cham, Switzerland, pp. 241–270.

Sun Q., Ye Z.H., Wang X.R. and Wong M.H. 2005. Increase of glutathione in mine population of *Sedum alfredii*: A Zn hyperaccumulator and Pb accumulator. *Phytochemistry* 66: 2549–2556.

Sun Y., Zhou Q., Wang L. and Liu W. 2009. Cadmium tolerance and accumulation characteristics of *Bidens pilosa* L. as a potential Cd-hyperaccumulator. *J. Hazard. Mater* 161: 808–814.

Tang Y.-T., Qiu R.-L., Zeng X.-W., Ying R.-R., Yu F.-M. and Zhou X.-Y. 2009. Lead, zinc cadmium accumulation and growth simulation in *Arabis paniculata* Franch. *Environ Exp. Bot.* 66: 126–134.

Tavallali V., Rahemi M., Eshghi S., Kholdebarin B. and Ramezanian A. 2010. Zinc alleviates salt stress and increases antioxidant enzyme activity in the leaves of pistachio (*Pistacia vera* L. 'Badami') seedlings. *Turk. J. Agr. Forest* 34: 349–359.

Thomas F., Malick C., Endreszl E.C. and Davies K.S. 1998. Distinct responses to copper stress in the halophyte, *Mesembryan-themum crystallium*. *Physiol. Plantarum* 102: 360–368.

Tisdale S.L., Nelson W.L. and Beaten J.D. 1984. *Zinc in Soil Fertility and Fertilizers*, 4th ed. Macmillan Publishing Company, New York.

Tremel A., Masson P., Sterckeman T., Baize D. and Mench M. 1997. Thallium in French agrosystems. I. Thallium content in arable soils. *Environ. Pollut.* 96: 293–302.

van der Ent A., Baker A.J., Reeves R.D., Chaney R.L., Anderson C.W., Meech J.A., Erskine P.D. et al. 2015b. Agromining: Farming for metals in the future? *Environ. Sci. Technol.* 49: 4773–4780.

van der Ent A., Erskine P. and Sumail S. 2015a. Ecology of nickel hyperaccumulator plants from ultramafic soils in Sabah (Malaysia). *Chemoecology* 25: 243–259.

Wang H. and Zhong G. 2011. Effect of organic ligands on accumulation of copper in hyperaccumulator and nonaccumulator *Commelina communis*. *Biol. Trace Elem. Res.* 143: 489–499.

Wild H. 1974. Indigenous plants and chromium in Rhodesia. *Kirkia* 9: 233–241.

Wilson-Corral V. 2008. Hiperacumulación de oro inducida químicamente en ocho especies vegetales (*Brassica juncea, Brassica campestris, Helianthus annus, Amaranthus spp., Sesamum indicum, Sorghum halepense, Amoreuxia palmatifida y Gossypium hirsutum*). PhD Thesis. Centro de Estudios Justo Sierra (CEJUS).

Xue S.G., Chen Y.X., Reeves R.D., Baker A.J., Lin Q. and Fernando D.R. 2004. Manganese uptake and accumulation by the hyperaccumulator plant *Phytolacca acinosa* Roxb. (Phytolaccaceae). *Environ. Pollut.* 131: 393–399.

Zhang X.H., Liu J., Huang H.T., Chen J., Zhu Y.N. and Wang D.Q. 2007. Chromium accumulation by the hyperaccumulator plant *Leersia hexandra* Swartz. *Chemosphere* 67: 1138–1143.

Zhao F.J., Lombi E., Breedon T. and McGrath S.P. 2000. Zinc hyperaccumulation and cellular distribution in *Arabidopsis halleri*. *Plant Cell Environ.* 23: 507–514.

10 MicroRNAs
Biogenesis and Therapeutic Functions

Atul Bhargava and Shilpi Srivastava

CONTENTS

10.1 INTRODUCTION

MicroRNAs (miRNAs) are a family of 21- to 25-nucleotide-long RNAs expressed in a wide variety of organisms. These act as novel agents exercising posttranscriptional control over most eukaryotic genomes. MicroRNAs account for approximately more than 3% of all human genes, and many miRNAs are highly conserved across species. The elements of the miRNA have been found in archaeabacteria and eubacteria, and this reveals their primitive ancestry. Although 100–200 miRNAs are known to be expressed in lower metazoans, 1000 or more miRNAs are predicted to regulate about 30% of the human genes. The number of miRNAs discovered in different organisms is bound to increase with the accessibility and usage of high-throughput sequencing. MicroRNAs originate from precursor RNAs that are transcribed from genomic DNA and target messenger RNAs (mRNAs) through hybridization to their 3'-untranslated regions (3' UTRs). This results in the breakdown of RNA and/or repression of translational repression, leading to changes in gene expression.

10.2   DISCOVERY AND ABUNDANCE

The first miRNA, *lin-4*, was discovered in *Caenorhabditis elegans* by the joint efforts of Victor Ambros's and Gary Ruvkun's laboratories. *Caenorhabditis elegans* has four different larval stages (L1–L4), and the gene *lin-4* is known to control the timing of postembryonic development in *C. elegans*, its activity being required for the transition from L1 to L2 stage of larval development. Mutations in *lin-4* interfere with the temporal regulation of larval development, causing irregularity in the cell division pattern of L1 (the first larval stage). Animals with *lin-4* loss-of-function mutations reported deficiency in adult structures, display inability to lay eggs, and exhibit discrepancy in the programming of larval stages. Fergunsson et al. in 1987 found that a suppressor mutation in the gene *lin-14* led to the development of a different phenotype of the *null-lin-4* mutations and in fact reverted the *null-lin-4* mutation phenotype. The results indicated that *lin-4* could negatively regulate *lin-14*. Ambros, together with Rosalind Lee and Rhonda Feinbaum, introduced mutations in the putative open reading frames (ORFs) and deduced that *lin-4* did not encode a protein but only two very small *lin-4* transcripts of 61 and 22 nucleotides in length. At the same time, Ruvkun, along with Bruce Wightman and Ilho Ha, discovered that *lin-14* was downregulated at a posttranscriptional level, and the 3′ UTR region of *lin-14* was sufficient for the temporal regulation. Ambros and Ruvkun while working independently concluded that the small and non-protein-coding transcript *lin-4* regulates *lin-14* through its 3′ UTR region. The negative regulation of *lin-14* protein expression requires an intact 3′ UTR of its mRNA, along with a functional *lin-4* gene. The shorter *lin-4* RNA later came into prominence as a member of an abundant class of tiny regulatory RNAs known as microRNAs or miRNAs. In the year 2000, after 7 years of the initial identification of *lin-4*, the second miRNA *let-7* was discovered, which encoded a temporally regulated 21-nucleotide small RNA that controlled the transition from the L4 stage into the adult stage. The identification of *let-7* was another evidence of developmental regulation by small RNAs and led to the supposition that such RNAs might be present in other animal and plant species as well. An important observation was that *let-7* sequence was found to be conserved across species from flies to humans, with the *let-7* RNA being detected in ascidians, annelids, arthropods, mollusks, hemichordates, and vertebrates. Subsequently, miRNAs were identified in many multicellular organisms, including flowering plants, worms, flies, fish, frogs, and mammals, as well as in single-cellular algae and DNA viruses.

10.3   BIOGENESIS

MicroRNAs are transcribed as precursor RNAs from intergenic, intronic, or polycistronic genomic loci by the enzyme RNA polymerase II (Pol II) in a multistage process that involves both the nuclear and cytoplasmic compartments (Figure 10.1). The primary miRNA transcript forms a stem-loop structure, which is recognized and processed in the nucleus by the Drosha and DiGeorge syndrome critical region 8 (DGCR8; also known as Pasha in flies and nematodes) RNase III complex, commonly termed the spliceosome apparatus.

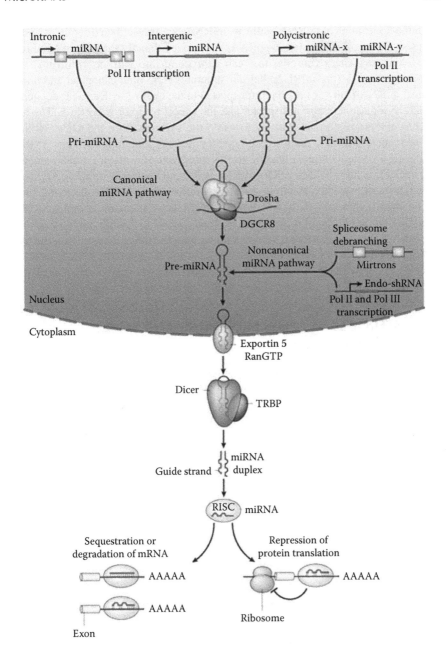

FIGURE 10.1 The biogenesis pathway of miRNA. (Reprinted by permission from Macmillan Publishers Ltd. *Nat. Rev. Mol. Cell Bio.*, Rottiers, V. and Näär, A.M., 2012, copyright 2012.)

In the noncanonical miRNA pathway, miRNAs are transcribed directly as endogenous short hairpin RNAs (endo-shRNAs) or derived directly through splicing from intervening sequences (introns) that can restructure into hairpins (mirtrons). The trimmed precursor (pre-miRNA) hairpin structures from both canonical and noncanonical miRNA pathways are then transported to the cytosol by RanGTP-dependent nucleo-/cytoplasmic cargo transporter and the export receptor exportin 5. The hairpin precursors are further processed inside the cytoplasm by the Dicer (RNAse III endonuclease) and transactivation-response RNA-binding protein (TRBP) RNase III enzyme complex, leading to the formation of a mature double-stranded ~22-nucleotide miRNA. MicroRNA duplex is then unwinded by utilizing Argonaute (AGO) proteins, which also aid in the incorporation of the miRNA-targeting strand (guide strand) into the AGO-containing RNA-induced silencing complex (RISC). The RISC–miRNA complex so formed is then guided to specific target sequences in mRNAs. The RISC–miRNA complex recognizes the mRNAs mainly on the basis of Watson–Crick base pairing of nucleotides 2–8 in the mature miRNA (termed the seed sequence), with particular mRNA sequences mainly located on the 3′ UTR. The additional base pairing affords greater affinity and targeting efficiency.

10.4 AFFINITIES BETWEEN MICRORNAs AND SMALL INTERFERING RNAs

Small interfering RNAs (siRNAs) are single-stranded RNA molecules (normally 21–25 nucleotides long) that are produced by the cleavage and processing of double-stranded RNA. The siRNAs bind to complementary sequences present in mRNA and bring about cleavage and degradation of the mRNA. According to He and Hannon (2004), siRNAs and miRNAs are similar in their molecular nature, biogenesis, and functions. Both miRNAs and siRNAs are small RNAs of 18–25 nucleotides in length and share a common RNase-III processing enzyme, Dicer, and closely related effector complexes, RISCs, for posttranscriptional repression. However, miRNA and siRNA are quite different, and the major differences are depicted in Table 10.1.

TABLE 10.1
Major Differences between miRNA and siRNA

miRNA	siRNA
Derived from genomic loci distinct from other recognized genes	Derived from mRNAs, transposons, viruses, or heterochromatic DNA
Processed from transcripts that can form local RNA hairpin structures	Processed from long bimolecular RNA duplexes or extended hairpins
A single miRNA:miRNA duplex is generated from each miRNA hairpin precursor molecule	A multitude of siRNA duplexes is generated from each siRNA precursor molecule, leading to many different siRNAs accumulating from both strands of this extended double-stranded RNA
miRNA sequences are nearly always conserved in related organisms	These are rarely conserved

10.5 FUNCTIONS OF MICRORNAs

In invertebrates, miRNAs regulate developmental timing, control of growth, neuronal differentiation, cell proliferation, and programmed cell death. In humans, miRNAs seem to be involved in almost all biological processes that include cellular growth and development, differentiation, embryogenesis, stem cell maintenance, autophagy, hematopoietic cell differentiation, brain development, proliferation, and apoptosis. Despite this, the biological role and *in vivo* functions of most mammalian miRNAs are still poorly understood.

10.6 THERAPEUTIC APPLICATIONS

10.6.1 ROLE IN CARDIOVASCULAR DISEASES

Cardiovascular diseases (CVDs), especially coronary artery diseases (CADs), are the major cause of death in many parts of the world. In recent years, a large number of data suggest that miRNAs play an important role in the development of CVDs. A huge amount of data at our disposal points toward the role of miRNAs as promising new diagnostic and prognostic biomarkers in CVDs, in addition to established protein-based biomarkers. Identifying an miRNA signature is therefore an essential prerequisite to studying the biological functions of these molecules in CVDs. MicroRNAs are considered disease-specific biomarkers, since specific circulating miRNAs are altered in various heart-related disorders such as myocardial infarction (MI), atherosclerosis, CAD, heart failure, atrial fibrillation, hypertrophy, and fibrosis.

Recent literature points to the role of miRNAs in cardiac myocyte growth. Hypertrophic growth of cultured cardiomyocytes has been induced by the overexpression of miR-23a, miR-23b, miR-24, miR-195, or miR-214 via adenovirus-mediated gene transfer, whereas overexpression of miR-150 or miR-181b caused a decrease in the size of cardiomyocyte cells. Likewise, the role of miRNAs in the development of heart has been well documented in several studies. The pivotal role of miRNAs in their development of CVDs has been indirectly demonstrated in Dicer-deficient mice that lose miRNAs, resulting in the impairment of both heart and vessel development. miRNAs also play a major role in angiogenesis which is an important physiological and pathological process. Studies in Dicer-deficient mice have shown severe reduction in blood vessel formation that points to the fact that miRNAs may play a crucial role in angiogenesis in the mammalian class. It has been documented that miRNAs have significant role in cardiac hypertrophy and heart failure, since miRNAs regulate the differentiation and growth of cardiac cells. It has been proved that the overexpression of some miRNAs that are upregulated in hypertrophic hearts induces cardiac myocyte hypertrophy, whereas the overexpression of other miRNAs that are downregulated prevents the occurrence of cardiac myocyte hypertrophy. Several clinical studies have pointed toward the role of miRNAs in human cardiac hypertrophy and heart failure. As of now, the use of miRNA mimics and antagomirs, with specific reference to cardiovascular research, has not percolated down to the clinical trials, but encouraging results reflect a pivotal role of miRNA therapeutics in CVDs in the decades to come.

10.6.2   ROLE IN NEURODEGENERATIVE DISEASES

Neurodegenerative diseases are a diverse class of disorders that involve progressive degeneration of the central nervous system or the peripheral nervous system. Some common neurodegenerative disorders include Parkinson's disease (PD) and Alzheimer's disease, whereas rare disorders include Huntington's disease, Creutzfeldt–Jakob disease, spinocerebellar ataxia, transmissible spongiform encephalopathies, frontotemporal dementia, Benson's syndrome, and amyotrophic lateral sclerosis. Positron emission tomography has shown that human neurodegenerative disorders are chronic conditions whose underlying pathological processes last for decades rather than just for years. Most of the neurodegenerative disorders share some common features such as their appearance during the later stages of life, extensive loss of neurons, synaptic abnormalities, and the presence of cerebral deposits of misfolded protein aggregates, which are a typical disease signature. The neurodegenerative diseases involve a gradual progression of pathological conditions in a pattern that collectively leads to neuronal death and/or compromised connectivity. Many neurodegenerative diseases are caused by genetic mutations, and neurodegenerative brains show evidence of considerable cellular stress. In some neurodegenerative disorders, brain RNAs become pathologically active and show unusual pattern of RNA oxidation, variations in RNA splicing, and changes in ribosomes that cause abnormalities in mRNA translations, mainly due to frameshift mutations.

MicroRNAs are enriched in human brain, and more than 1000 miRNAs are known to be expressed in human brain. The expression of brain miRNAs significantly changes during the development of the brain. Some miRNAs are relatively enriched during early development in the mammalian brain, whereas others are enriched during the later stages of development. It has been conclusively established that changes in miRNA networks in the brain play a significant role in the development of neurodegenerative disorders. MicroRNAs in the brain play an important role in cell fate determination along developmental lineages, in neuroplasticity, and in many other neurobiological functions. The expressions of miRNAs are known to change drastically in response to various abiotic stresses such as cold, heat, hypoxia, irradiation, and nutritional scarcity, with a subset of miRNAs getting increased and another being decreased. For example, miR-93 is known to be upregulated in titanium-induced stress in a human osteoblast cell line and in the liver of tamoxifen-fed rats but is downregulated in hypoxic human trophoblasts. Thus, miRNA expression patterns are likely to change and are expected to modify the translation of a large number of cellular mRNAs during conditions of cellular stress such as neurodegenerative diseases (Table 10.2).

Neural cell death is the characteristic feature of most neurodegenerative disorders and the principal cause of many functional deficits. Another common feature among most of the neurodegenerative disorders is the accumulation of proteins that are toxic to neurons. The accumulation of these toxic proteins can be modulated by miRNAs either by regulating the mRNA encoding the toxic protein or by regulating the mRNAs encoding proteins that modulate the expression of the disease-causing protein (Figure 10.2). Apart from this, miRNAs also contribute to the progressive development of neurodegenerative disease downstream of the accumulation of toxic proteins by changes in the expression of polypeptides that promote or inhibit cell survival.

TABLE 10.2
Human MicroRNAs Linked to Neurodegenerative Disease

MicroRNA	Mode of Action	Targets	References
miR-19, miR-101, miR-130	Suppresses ATAXIN1 accumulation	ATAXIN1	Lee et al. (2008)
miR-9/miR-9*	Suppresses negative interaction between huntingtin and REST/CoREST	REST and CoREST	Packer et al. (2008)
miR-29a/b	Suppresses the accumulation of toxic Aβ peptide	BACE1	Hebert et al. (2008)
miR-133b	Dopamine neuron specification and survival (?)	PITX3	Kim et al. (2007)
miR-433	Indirectly suppresses the expression of α-synuclein	FGF20	Wang et al. (2008)
miR-659	Represses GRN expression	GRN	Rademakers et al. (2008)

Source: Reprinted by permission from Macmillan Publishers Ltd. *Nat. Rev. Neurosci.*, Eacker, S.M. et al., 2009, copyright 2009.

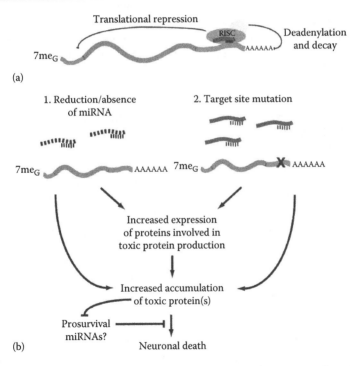

FIGURE 10.2 Messenger RNA repression by miRNA and its effect on neurodegeneration. (a) MicroRNAs (dark gray) bind their target mRNAs through sequences in the 3′ UTR. (b) Proposed mechanisms by which miRNAs could influence neurodegeneration. (Reprinted by permission from Macmillan Publishers Ltd. *Nat. Rev. Neurosci.*, Eacker, S.M. et al., 2009, copyright 2009.)

Alzheimer's disease, named after the German psychiatrist Alois Alzheimer, is an irreversible, progressive brain disorder that usually starts slowly and aggravates over time, destroying the memory and thinking skills. Alzheimer's disease is a protein-misfolding disease caused by plaque accumulation of abnormally folded amyloid beta protein and tau protein in the affected brain. Studies in Alzheimer's disease have revealed that several members of the miR-16 family (miR-16, miR-15, miR-195, and miR-497) could regulate endogenous ERK1 and tau phosphorylation in neurons. ERK1, a Tau kinase, is directly targeted by miR-15, and therefore, modification of miR-15 is considered the root cause of the abnormal tau phosphorylation observed in Alzheimer's disease. Increased levels of β-amyloid precursor protein (APP) and reduced levels of miR-153 have been observed in the brains of patients with Alzheimer's disease. Hebert et al. (2008) performed expression profiling of 328 miRNAs in patients with Alzheimer's disease and showed a decrease in the expression of 13 miRNAs in the diseased patients. The miRNAs miR-29a, miR-29b-1, and miR-9 significantly decreased the expression of β-amyloid cleavage enzyme 1 (BACE-1), a critical enzyme that is required for the cleavage of APP and generation of toxic β-amyloid species.

Parkinson's disease is a long-term degenerative disorder of the central nervous system that involves the death of nerve cells in the brain. Tremors, rigidity, bradykinesia, and postural instability are the common symptoms that generally come slowly over time and typically occur in people older than 60 years of age. Expression profiling of 224 miRNAs obtained from the brain of patients inflicted with PD showed distinct variations in several miRNAs, including miR-133b. Recent miRNA profiling studies of brains of patients with PD have revealed a decreased expression of miR-34b and miR-34c in the affected areas of the brain such as frontal cortex, amygdala, substantia nigra, and cerebellum. A further knockdown of miR-34b/34c in differentiated SH-SY5Y neuroblastoma cells showed changes in mitochondrial activity and oxidative stress and associated cell death, which are the major biochemical abnormalities observed in PD.

Huntington's disease is a common hereditary neurodegenerative disorder that causes movement, psychiatric, and cognitive (loss of thinking ability) symptoms due to an autosomal dominant mutation. The symptoms usually become noticeable between the ages of 35 years and 44 years. Expression profiling of miRNAs in the cerebral cortex of HD-afflicted patients showed a significant decrease of miR-9 and miR-9* with the progression of the disease. Variations in miR-9 and miR-9* can affect the expression of the RE-1 silencing transcription factor (REST) and its corepressor protein, CoREST. It has been noticed that the upregulation of the REST repressor complex could have harmful effects on gene expression of the neurons and can suppress the prosurvival genes such as *BDNF* that may significantly contribute to the development of HD. Another miRNA miR-132 has been found to be downregulated in the brains of HD mouse models and in the postmortem brains of patients with HD. As a result of this decrease, its mRNA target p250GAP, which encodes a member of the Rac/Rho family of GTPase-activating proteins, is enhanced. Thus, further studies are required for the identification of additional mRNAs targets of miR-9/miR-9* and miR-132, which can lead to a greater understanding of the molecular pathogenesis of HD.

10.6.3 ROLE IN CANCER THERAPEUTICS

Cancer is a consequence of disordered gene expression in which there is a breakdown of gene regulatory networks that maintain a balance between oncogenes and tumor suppressor genes. Cancer is a major human health problem around the world and is the leading cause of death in many parts of the world. More than 60% of the world's new cancer cases occur in Africa, Asia, and Central and South America, and about 70% of the deaths due to cancer also occur in these regions. In 2012, there were 14 million new cases and 8.2 million cancer-related deaths around the globe. Most of the deaths occurring due to cancer take place in underdeveloped and developing countries. Deaths from cancer in the world are continuously rising, with around 9 million people dying from cancer in 2015 and about 11.4 million in 2030. Cancer is associated with an abnormal increase in the number of cells and variations in mechanisms that regulate birth of new cells and their proliferation. Cancer is unique in comparison with other tumor-forming processes, owing to its ability to invade the surrounding fresh tissues. Cancer development is a multistep process during which normal cells gradually acquire abnormal proliferative and invasive properties. Tumorigenesis is a unique form of natural selection process that involves the accumulation of multiple somatic mutations in populations that are in the process of undergoing neoplastic transformation. Numerous exogenous chemical, physical, or biological factors known as carcinogens are known to induce cancer. The carcinogens act on human beings, who vary in their ability to cope with them due to genetic, psychological, social, economic variations, as well as due to endogenous processes in the human body that also contribute to the development of cancer. The exogenous carcinogens are usually grouped into physical, chemical, and biological agents.

Several forms of molecular alterations such as gene amplifications, insertions, deletions, rearrangements, and point mutations have been documented in human cancers. A number of positive and negative mediators of cell growth and differentiation have been established and characterized that define the basic role for these genetic elements in neoplastic transformation and tumor development. It has been amply proven by microarray-based gene expression studies that cancer is a disorder involving abnormal gene expression. Somatic mutations that occur during the development of cancers alter the gene expression patterns that result in significant changes in cellular physiology, including abnormally regulated cell proliferation, and acquisition of invasive behavior. However, the gene expression signature of a specific cancer can be positively utilized in the diagnosis and prediction of responses to therapy.

MicroRNAs have often been located in fragile regions of the chromosome that are susceptible to structural rearrangements, deletions, and amplification and therefore have been implicated in malignancy. All the available evidence, especially the expression profiling data, points toward nearly ubiquitous dysregulation of miRNA expression in human cancers.

The rationale for the role of miRNA as a tumor suppressor or oncogene is supported by different types of evidence, provided as follows:

1. Widespread dysregulation of miRNA expression in diverse cancers
2. Loss or gain of miRNA function in tumor formation, owing to mutation, deletion, or amplification

3. Direct evidence of tumor-promoting or tumor-suppressing activity, using animal models
4. Detailed information of cancer-relevant targets that throw light on how miRNA participates in oncogenesis

The first report of the involvement of miRNA in cancer development appeared in 2002 during a study on the deletion in chromosome 13, which is the most frequent chromosomal abnormality in chronic lymphocytic leukemia (CLL). Calin et al. (2002) found two miRNA genes, mir-15 and mir-16, located within this 30-kb deletion and whose miRNAs were absent or downregulated in the majority (68%) of cases. It was suggested that these two miRNAs were somehow involved in the development of CLL. It has now been generally recognized that global miRNA loss enhances tumorigenesis. Similarly, several later studies showed that miR-34a was a direct transcriptional target of the p53 tumor suppressor protein, commonly known as the *cellular gatekeeper*, which is a critical regulator during cellular stress.

Considering their potential implications in tumor development, numerous efforts have centered around the miRNA profiling expression in different types of cancers such as lymphoma, lung cancer, colorectal cancer, myeloid and lymphoid leukemias, breast cancer, prostate cancer, ovarian cancer, and testicular germ cell cancer. MicroRNA expression profiles in cancerous condition provide important insights for a deeper understanding of the tumor development or metastases as well as for cancer diagnosis and prognosis. The expression profile of miRNA in cancerous condition seems to be specific for a particular type of tissue, since different types of tumors have distinctive patterns of miRNA expression. For example, unique miRNA signatures that distinguish between benign tumors and carcinoma tumors have been established for prostate cancer. A number of miRNAs are upregulated, whereas several others are downregulated during tumorigenesis.

Several mechanisms have been put forward with respect to the abnormal expression of miRNAs during cancer:

1. The discovery of miRNA loci at chromosomal breakpoint and specific genomic regions associated with cancer
2. The expression of the miRNA can be influenced by the regulation of epigenetic silencing such as methylation and histone-modification losses
3. The failure of the Drosha-processing step leads to widespread downregulation of miRNAs
4. The processing of miRNA can be influenced by point mutation in the miRNA precursor
5. Impairing of miRNA function by the introduction of mutation in a target sequence

10.6.4 Role in Autoimmune Disorders

Regulation of the immune system is of prime importance in the prevention of several pathogenic disorders, including autoimmune disease. MicroRNAs influence the development and regulation of the adaptive immune system and play an important role in modulating innate immune responses against a variety of pathogens, including viruses and bacteria. Autoimmune disorders or diseases originate from the deficiency of immunological tolerance to autoantigens and resultant pathological status, both of which affect specific target organs or multiple organ systems. Some common autoimmune diseases include celiac disease, psoriasis, rheumatoid arthritis, primary biliary cholangitis, diabetes mellitus type 1, Graves' disease, multiple sclerosis, inflammatory bowel disease, idiopathic thrombocytopenic purpura (or autoimmune thrombocytopenic purpura), and systemic lupus erythematosus. Autoimmune diseases start during the adulthood, with women being more commonly affected than men.

Recent researches have indicated that miRNAs play an important role in the regulation of immunological functions and the prevention of autoimmune disorders. The miRNA miR-181a, overexpressed in the thymus cells but expressed at much lower levels in organs such as heart, lymph nodes, and bone marrow, was one of the first miRNAs reported to have a major role in immune cell development. The roles of miRNAs in different autoimmune disorders have been provided in Table 10.3.

TABLE 10.3
Role of miRNAs in Autoimmune Disorders

miRNA	Autoimmune Disorder	Upregulation/Downregulation
miR-146a	Rheumatoid arthritis	Upregulated
miR-155	Rheumatoid arthritis	Upregulated
miR-26	Rheumatoid arthritis	Upregulated
miR-34a*	Rheumatoid arthritis	Downregulated
miR-15a	Rheumatoid arthritis	Downregulated
miR-124a	Rheumatoid arthritis	Downregulated
miR-141	Systemic lupus erythematosus	Downregulated
miR-383	Systemic lupus erythematosus	Downregulated
miR-299-3p	Systemic lupus erythematosus	Upregulated
miR-146	Sjögren's syndrome	Upregulated
miR-18b	Multiple sclerosis	Upregulated
miR-493	Multiple sclerosis	Upregulated
miR-599	Multiple sclerosis	Upregulated
miR-17-5p	Multiple sclerosis	Upregulated
miR-497	Multiple sclerosis	Downregulated
miR-126	Multiple sclerosis	Downregulated
miR-155	Inflammatory bowel disease	Upregulated
miR-126	Inflammatory bowel disease	Upregulated
miR-31	Inflammatory bowel disease	Upregulated
miR-200b	Inflammatory bowel disease	Downregulated
miR-320a	Inflammatory bowel disease	Downregulated
miR-26	Inflammatory bowel disease	Downregulated

10.7 MICRORNAs AND PHARMACOGENOMICS

Pharmacogenomics is a modern branch of science that focuses on the identification of genetic variants that influence the effect of drug via alterations in pharmacokinetics. It delves deep into the absorbtion, distribution, metabolization, and elimination of drugs or in pharmacodynamics by modifying their target or reorientation of the biological pathways that shape a patient's sensitivity to the pharmacological effects of drugs. The current focus is on studying the phenotypic effects of single-nucleotide polymorphisms (SNPs) and other genetic variations in candidate drug-metabolization and drug-target genes. In recent years, SNPs have been utilized to understand the susceptibility to specific diseases in the human population. MicroRNAs are at the center of studies for determining how patients react to and metabolize drugs, since the drug metabolism can involve a group of genes. MicroRNA pharmacogenomics is the study of miRNAs and polymorphisms affecting miRNA function for predicting the behavior of drug and improving its efficacy. Shomron (2010) provided a strong association between miRNA and pharmacogenomics by discussing several ways in which miRNAs can modulate the expression of genes and how the gene product affects an individual's response to drugs. Since the miRNA-binding site is present in a gene's 3′ UTR, at least one of the following conditions should be fulfilled:

1. A change in the levels of cellular miRNA
2. Inherited modifications in the miRNA-binding site on the 3′ UTR such as SNPs (or de novo generation of an miRNA-binding site)
3. Sequence modification in the miRNA gene that affects its biogenesis, expression, stability, or binding affinities to the target gene

Recent researches have suggested that changes in the sequence of a miRNA and/or variations in the target region of an miRNA-regulated transcript can have major effects in posttranscriptional regulation of proteins. Protein diversity as a result of alternative splicing might affect important genes in multiple pathways associated with the activation of prodrugs and their metabolism. It is also interesting to note that the variations in nucleotide sequence such as SNPs can influence the way in which miRNAs regulate their targets, indicating their active role in drug metabolism and phenotypic variation. In fact, the term miRSNP/miR-polymorphism has been introduced as a novel class of SNPs/polymorphisms that interfere with the functioning of miRNA. Since miRNAs are attractive drug targets and regulate expression of several important proteins in the cell miRNA, pharmacogenomics is bound to have strong clinical implications. Polymorphism in miRNAs can hamper the functioning of miRNA, resulting in loss of the miRNA-mediated regulation of a drug-target gene that is likely to confer drug resistance. Therefore, the miR-polymorphisms can be utilized as predictors of drug response and are likely to result in the development of reliable methods of determining appropriate drug dosages based on a patient's genetic constitution, which would decrease the likelihood of a drug overdose. Therefore, miR-polymorphisms show tremendous scope for use in disease prognosis and diagnosis, and deciphering the role and functions of miRNA-polymorphisms has excellent prospects in pharmacogenomics, molecular epidemiology, and personalized medicine.

BIBLIOGRAPHY

Alvarez-Garcia I. and Miska E.A. 2005. MicroRNA functions in animal development and human disease. *Development* 132: 4653–4662.

Ambros V. 2004. The functions of animal microRNAs. *Nature* 431: 350–355.

Barbato C., Ruberti F. and Cogoni C. 2009. Searching for MIND: MicroRNAs in neurodegenerative diseases. *J. Biomed. Biotechnol.* 2009: 871313.

Bartel D.P. 2004. MicroRNAs: Genomics, biogenesis, mechanism, and function. *Cell* 116: 281–297.

Bartel D.P. 2009. MicroRNAs: Target recognition and regulatory functions. *Cell* 136: 215–233.

Barwari T., Joshi A. and Mayr M. 2016. MicroRNAs in cardiovascular disease. *J. Am. Coll. Cardiol.* 68: 2577–2584.

Betel D., Wilson M., Gabow A., Marks D.S. and Sander C. 2008. The microRNA.org resource: Targets and expression. *Nucl. Acids Res.* 36: D149–D153.

Bilen J., Liu N. and Bonini N.M. 2006. A new role for microRNA pathways: Modulation of degeneration induced by pathogenic human disease proteins. *Cell Cycle* 5: 2835–2838.

Bilen J., Liu N., Burnett B.G., Pittman R.N. and Bonini N.M. 2006. MicroRNA pathways modulate polyglutamine-induced neurodegeneration. *Mol. Cell* 24: 157–163.

Bishop J.M. 1991. Molecular themes in oncogenesis. *Cell* 64: 235–248.

Bothwell M. and Giniger E. 2000. Alzheimer's disease: Neurodevelopment converges with neurodegeneration. *Cell* 102: 271–273.

Bruijn L.I., Miller T.M. and Cleveland D.W. 2004. Unraveling the mechanisms involved in motor neuron degeneration in ALS. *Annual Rev. Neurosci.* 27: 723–749.

Buckland J. 2010. Biomarkers: MicroRNAs under the spotlight in inflammatory arthritis. *Nat. Rev. Rheumatol.* 6: 436.

Caldas C. and Brenton J.D. 2005. Sizing up miRNAs as cancer genes. *Nat. Med.* 11: 712–714.

Calin G.A. and Croce C.M. 2006. MicroRNA signatures in human cancers. *Nat. Rev. Cancer* 6: 857–866.

Calin G.A., Dumitru C.D., Shimizu M., Bichi R., Zupo S., Noch E., Aldler H. et al. 2002. Frequent deletions and down-regulation of micro-RNA genes miR15 and miR16 at 13q14 in chronic lymphocytic leukemia. *Proc. Natl. Acad. Sci.* (*USA*) 99: 15524–15529.

Chen X., Li X., Guo J., Zhang P. and Zeng W. 2017. The roles of microRNAs in regulation of mammalian spermatogenesis. *J. Anim. Sci. Biotechnol.* 8: 35.

Christensen M. and Schratt G.M. 2009. MicroRNA involvement in developmental and functional aspects of the nervous system and in neurological diseases. *Neurosci. Lett.* 466: 55–62.

Chung I.-M., Ketharnathan S., Thiruvengadam M. and Rajakumar G. 2016. Rheumatoid arthritis: The stride from research to clinical practice. *Int. J. Mol. Sci.* 17: 900.

Coleman W.B. and Tsongalis G.J. 2006. Molecular mechanisms of human carcinogenesis. In: Bignold L.P. (Ed.). *Cancer: Cell Structures, Carcinogens and Genomic Instability.* Birkhäuser Verlag, Basel, Switzerland.

Coppede F., Mancuso M., Siciliano G., Migliore L. and Murri L. 2006. Genes and the environment in neurodegeneration. *Biosci. Rep.* 26: 341–367.

Eacker S.M., Dawson T.M. and Dawson V.L. 2009. Understanding microRNAs in neurodegeneration. *Nat. Rev. Neurosci.* 10: 837–841.

Fiore R., Siegel G. and Schratt G. 2008. MicroRNA function in neuronal development, plasticity and disease. *Biochim. Biophys. Acta* 1779: 471–478.

Garzon R., Calin G.A. and Croce C.M. 2009. MicroRNAs in cancer. *Annu. Rev. Med.* 60: 167–179.

Gidlöf O., Smith J.G., Miyazu K., Gilje P., Spencer A., Blomquist S. and Erlinge D. 2013. Circulating cardio-enriched microRNAs are associated with long-term prognosis following myocardial infarction. *BMC Cardiovasc. Disord.* 13: 12.

Gozuacik D., Akkoc Y., Ozturk D.G. and Kocak M. 2017. Autophagy-regulating microRNAs and Cancer. *Front. Oncol.* 7: 65.

He L. and Hannon G.J. 2004. MicroRNAs: Small RNAs with a big role in gene regulation. *Nat. Rev. Genet.* 5: 522–531.

Hébert S.S. and De Strooper B. 2009. Alterations of the microRNA network cause neurodegenerative disease. *Trends Neurosci.* 32: 199–206.

Hebert S.S., Horré K., Nicolaï L., Papadopoulou A.S., Mandemakers W., Silahtaroglu A.N., Kauppinen S., Delacourte A. and De Strooper B. 2008. Loss of microRNA cluster miR-29a/b-1 in sporadic Alzheimer's disease correlates with increased BACE1/beta-secretase expression. *Proc. Natl. Acad. Sci. (USA)* 105: 6415–6420.

Hoelscher S.C., Doppler S.A., Dreßen M., Lahm H., Lange R. and Krane M. 2017. MicroRNAs: Pleiotropic players in congenital heart disease and regeneration. *J. Thorac. Dis.* 9(Suppl 1): S64–S81.

Kim J., Inoue K., Ishii J., Vanti W.B., Voronov S.V., Murchison E., Hannon G. and Abeliovich A. 2007. A MicroRNA feedback circuit in midbrain dopamine neurons. *Science* 317: 1220–1224.

Kosik K.S. and Krichevsky A.M. 2005. The elegance of the MicroRNAs: A neuronal perspective. *Neuron* 47: 779–782.

Koturbash I., Tolleson W.H., Guo L., Yu D., Chen S., Hong H., Mattes W. and Ning B. 2015. MicroRNAs as pharmacogenomic biomarkers for drug efficacy and drug safety assessment. *Biomark. Med.* 9: 1153–1176.

Lee R.C., Feinbaum R.L. and Ambros V. 1993. The *C. elegans* heterochronic gene lin-4 encodes small RNAs with antisense complementarity to lin-14. *Cell* 75: 843–854.

Lee Y., Samaco R.C., Gatchel J.R., Thaller C., Orr H.T. and Zoghbi H.Y. 2008. MiR-19, miR-101 and miR-130 co-regulate ATXN1 levels to potentially modulate SCA1 pathogenesis. *Nat. Neurosci.* 11: 1137–1139.

Leung A.K. and Sharp P.A. 2007. MicroRNAs: A safeguard against turmoil? *Cell* 130: 581–585.

Li C., Feng Y., Coukos G. and Zhang L. 2009. Therapeutic microRNA strategies in human cancer. *AAPS J.* 11: 747–757.

Li N., Long B., Han W., Yuan S. and Wang K. 2017. MicroRNAs: Important regulators of stem cells. *Stem Cell Res. Ther.* 8: 110.

Li M.-P., Hu Y.-D., Hu X.-L., Zhang Y.-J., Yang Y.-L., Jiang C., Tang J. and Chen X.-P. 2016. MiRNAs and miRNA polymorphisms modify drug response. *Int. J. Environ. Res. Public Health* 13: 1096.

Loosen S.H., Schueller F., Trautwein C., Roy S. and Roderburg C. 2017. Role of circulating microRNAs in liver diseases. *World J. Hepatol.* 9: 586–594.

Lu J., Getz G., Miska E.A., Alvarez-Saavedra E., Lamb J., Peck D., Sweet-Cordero A. et al. 2005. MicroRNA expression profiles classify human cancers. *Nature* 435: 834–838.

Lund E., Güttinger S., Calado A., Dahlberg J.E. and Kutay U. 2004. Nuclear export of microRNA precursors. *Science* 303: 95–98.

Marsit C.J., Eddy K. and Kelsey K.T. 2006. MicroRNA responses to cellular stress. *Cancer Res.* 66: 10843–10848.

Martello G., Rosato A., Ferrari F., Manfrin A., Cordenonsi M., Dupont S., Enzo E. et al. 2010. A microRNA targeting dicer for metastasis control. *Cell* 141: 1195–1207.

Martin J.B. 1999. Molecular basis of the neurodegenerative diseases. *N. Engl. J. Med.* 340: 1970–1980.

Menghini R., Stohr R. and Federici M. 2014. MicroRNAs in vascular aging and atherosclerosis. *Ageing Res. Rev.* 17: 68–78.

Mishra P.J. and Bertino J.R. 2009. MicroRNA polymorphisms: The future of pharmacogenomics, molecular epidemiology and individualized medicine. *Pharmacogenomics* 10: 399–416.

Miska E.A. 2005. How microRNAs control cell division, differentiation and death. *Curr. Opin. Genet. Dev.* 15: 563–568.

Nelson P., Kiriakidou M., Sharma A., Maniataki E. and Mourelatos Z. 2003. The microRNA world: Small is mighty. *Trends Biochem. Sci.* 28: 534–540.

Nelson P.T., Wang W.X. and Rajeev B.W. 2008. MicroRNAs (miRNAs) in neurodegenerative diseases. *Brain Pathol.* 18: 130–138.

Nicoloso M.S., Spizzo R., Shimizu M., Rossi S. and Calin G.A. 2009. MicroRNAs—the micro steering wheel of tumour metastases. *Nat. Rev. Cancer* 9: 293–302.

Nishiguchi T., Imanishi T. and Akasaka T. 2015. MicroRNAs and cardiovascular diseases. *Biomed Res. Int.* 2015: 682857.

Oliveira-Carvalho V., Carvalho V.O., Silva M.M., Guimarães G.V. and Bocchi E.A. 2012. MicroRNAs: A new paradigm in the treatment and diagnosis of heart failure? *Arq. Bras. Cardiol.* 98: 362–369.

Olson E.N. 2014. MicroRNAs as therapeutic targets and biomarkers of cardiovascular disease. *Sci. Transl. Med.* 6: 239ps3.

Packer A.N., Xing Y., Harper S.Q., Jones L. and Davidson B.L. 2008. The bifunctional microRNA miR-9/miR-9* regulates REST and CoREST and is downregulated in Huntington's disease. *J. Neurosci.* 28: 14341–14346.

Pauley K.M., Cha S. and Chan E.K. 2009. MicroRNA in autoimmunity and autoimmune diseases. *J. Autoimmun.* 32: 189–194.

Peter M.E. 2009. Let-7 and miR-200 microRNAs: Guardians against pluripotency and cancer progression. *Cell Cycle* 8: 843–852.

Pillai R.S. 2005. MicroRNA function: Multiple mechanisms for a tiny RNA? *RNA* 11: 1753–1761.

Qu Z., Li W. and Fu B. 2014. MicroRNAs in autoimmune disease. *Biomed Res. Int.* 2014: 8. doi:10.1155/2014/527895.

Rademakers R., Eriksen J.L., Baker M., Robinson T., Ahmed Z., Lincoln S.J., Finch N. et al. 2008. Common variation in the miR-659 binding-site of GRN is a major risk factor for TDP43-positive frontotemporal dementia. *Hum. Mol. Genet.* 17: 3631–3642.

Reinhart B.J., Weinstein E.G., Rhoades M.W., Bartel B. and Bartel D.P. 2002. MicroRNAs in plants. *Genes Dev.* 16: 1616–1626.

Relling M.V. and Evans W.E. 2015. Pharmacogenomics in the clinic. *Nature* 526: 343–350.

Rossbach M. 2010. Small non-coding RNAs as novel therapeutics. *Curr. Mol. Med.* 10: 361–368.

Rottiers V. and Näär A.M. 2012. MicroRNAs in metabolism and metabolic disorders. *Nat. Rev. Mol. Cell Biol.* 13: 239–250.

Ruddon R.W. 1995. *Cancer Biology*, 3rd ed. Oxford University Press, New York.

Sassen S., Miska E.A. and Caldas C. 2008. Micro-RNA implications for cancer. *Virchows Arch.* 452: 1–10.

Sayed D., Hong C., Chen I.Y., Lypowy J. and Abdellatif M. 2007. MicroRNAs play an essential role in the development of cardiac hypertrophy. *Circ. Res.* 100: 416–424.

Schulte C. and Zeller T. 2015. microRNA-based diagnostics and therapy in cardiovascular disease—Summing up the facts. *Cardiovasc. Diagn. Ther.* 5: 17–36.

Shomron N. 2010. MicroRNA and pharmacogenomics. *Pharmacogenomics* 11: 629–632.

Skroblin P. and Mayr M. 2014. Going long: Long non-coding RNAs as biomarkers. *Circ. Res.* 115: 607–609.

Smolen J.S., Aletaha D., Koeller M., Weisman M.H. and Emery P. 2007. New therapies for treatment of rheumatoid arthritis. *Lancet* 370: 1861–1874.

Soifer H.S., Rossi J.J. and Saetrom P. 2007. MicroRNAs in disease and potential therapeutic applications. *Mol. Ther.* 15: 2070–2079.

Soto C. 2003. Unfolding the role of protein misfolding in neurodegenerative diseases. *Nat. Rev. Neurosci.* 4: 49–60.

Urbich C., Kuehbacher A. and Dimmeler S. 2008. Role of microRNAs in vascular diseases, inflammation, and angiogenesis. *Cardiovasc. Res.* 79: 581–588.

van Rooij E., Marshall W.S. and Olson E.N. 2008. Toward microRNA-based therapeutics for heart disease: The sense in antisense. *Circ. Res.* 103: 919–928.

Voinnet O. 2009. Origin, biogenesis, and activity of plant microRNAs. *Cell* 136: 669–687.

Wang G., van der Walt J.M., Mayhew G., Li Y.J., Züchner S., Scott W.K., Martin E.R. and Vance J.M. 2008. Variation in the miRNA-433 binding site of FGF20 confers risk for Parkinson disease by overexpression of alpha-synuclein. *Am. J. Hum. Genet.* 82: 283–289.

Zhang B., Pan X., Cobb G.P. and Anderson T.A. 2007. MicroRNAs as oncogenes and tumor suppressors. *Dev. Biol.* 302: 1–12.

Zhang C. 2008. MicroRNAs: Role in cardiovascular biology and disease. *Clin. Sci.* 114: 699–706.

Index